RUBBER MIRROR

Henry J. Inman

RUBBER
MIRROR

REFLECTIONS OF THE

RUBBER DIVISION'S FIRST

100 YEARS

The University of Akron Press
Akron, Ohio

Copyright © 2009 by The Rubber Division of the American Chemical Society

First Edition 2009
Manufactured in the United States of America.
All inquiries and permission requests should be addressed to the Publisher,
The University of Akron Press, Akron, Ohio 44325–1703.

13 12 11 10 09 5 4 3 2 1

Library of Congress Cataloging-in-Publication Data

Inman, Henry J., 1946–
  Rubber mirror : reflections of the rubber division's first 100 Years /
Henry J. Inman. — 1st ed.
      p. cm.
  Includes bibliographical references and index.
  ISBN 978-1-931968-60-7 (cloth : alk. paper)
  1. Goodyear Tire and Rubber Company.   2. Rubber.   3. American Chemical
Society. Rubber Division.   4. American Chemical Society. Division of Rubber
Chemistry.   5. Rubber industry and trade—Ohio—Akron—History.   I. Title.
  TS1885.U62A45 2009
  678'.20973—dc22

                                                                2008055268

The paper used in this publication meets the minimum requirements of American
National Standard for Information Sciences—Permanence of Paper for
Printed Library Materials, ANSI z39.48–1984. ⊗

Cover design: Dave Szalay

CONTENTS

LIST OF ILLUSTRATIONS

For our children,
Gina (Jason), Michelle (Justin), Lara,
Rebecca, and Bryan, and our grandchildren Dylan,
Travis, Jack and Ayden.

PREFACE

People tend to forget that the word "history" contains the word "story."

KEN BURNS

It was late in September 1978 when I first moved to Akron, Ohio, from Oklahoma. Oklahoma! Remember? "Where the wind comes sweeping down the plains?" I arrived fresh from the newspaper business and was making the transition to the corporate world. A former college classmate and longtime friend, Gaylon White, encouraged me to move to Akron. He said it wasn't a bad place to live—in the summer. He didn't tell me about the ferocity and length of the winters in northeastern Ohio. I would have to experience that on my own. So partly out of naïveté, or just plain ignorance, I welcomed the challenge.

I took a deep breath when I got out of my car in the Seiberling Field parking lot of the Goodyear Tire & Rubber Company and received a pungent reminder of my surroundings. I thought to myself, What the hell is that smell? Is that burnt rubber? Coincidence never was far from me, and neither was a fellow employee. I asked him the same question that was bouncing in my brain, and I'll never forget his response. He laughed and said, "That's not rubber burning. That's the smell of money!" And so I received my introductory lesson in Rubber 101.

The last thirty years at Goodyear and in the rubber industry have been a great education for me, and there were many people who, whether they realized it or not, were mentors to me and played a role in the development of this book.

Working my way back in time from the most recent, Chris Laursen, the latest in the line of great librarians for the American Chemical Society's Rubber Division, provided invaluable help in my research. I could not have done it without him. "Goog" (a shortened version of Google, the online search engine) was the greatest help whenever I needed it. Equally valuable to my efforts was Ben Kastein, who in his early nineties is the sage of the Rubber Division. I beat a path between Cuyahoga Falls and his home in Silver Lake—it used to be called the Ohio Canal. Another great contributor was the legendary Ralph Graff. Thanks to him, I can write off hundreds of minutes of long distance phone calls. He was another great "go-to" guy.

The other Rubber Division people who were invaluable to my research efforts were Charles Rader, Melanie Avdeyev, Karen May, and, of course, Ed Miller, who not only shared some personal thoughts with me but also reviewed each chapter along the way. I also owe a great deal of gratitude to the many present and former Rubber Division members who willingly spent their valuable time with me to pass along their past knowledge. At the top of that list is Marge Bauer.

Contributors from the University of Akron also were numerous. Kent Marsden provided me with access to his personal thoughts and also served as a source of contacts within the university. Frank Kelly, Marsden's former boss, was a conduit of information, and Mark Bowles, author of *Chains of Opportunity* (University of Akron Press, 2008), gave me some great advice: "Just start writing!" Other University of Akron motivators were Kathleen Endres and David Ritchey, professors extraordinaire in the Department of Communication, and graphic designer Dave Szalay.

Other academics of note are Nancy Martin, University of Rochester archivist and Rochester collections librarian, and a University of Notre Dame undergraduate student, Bryan Fair, who will likely one day make his own mark in the world of the written word.

Several members of the Fourth Estate provided assistance, knowledge, and a critical eye to several chapters. They are Don Smith, editor of *Rubber World*; Bruce Meyer, managing editor of *Rubber & Plastics*

News; Ed Noga, editor of *Rubber & Plastics News*; Jack Sweeney, publisher of the *Houston Chronicle*; Cynthia Bates of the *Houston Chronicle*; and Dave Giffels and Steve Love, former *Akron Beacon Journal* writers and coauthors of *Wheels of Fortune* (University of Akron Press, 1999).

My appreciation also is extended to UK resident P. Graham Willis, retired Goodyear plantations expert, who reviewed the early chapters on rubber to ensure their accuracy. Another former Goodyear associate I mentioned earlier was Gaylon White. I can't mention him enough. He was a classmate of mine at the University of Oklahoma who was not only my mentor but also a very good friend and one the best writers I've ever known. Gaylon and I also owe a great deal of our enthusiasm of the written word to the late Harold Keith, sports information director emeritus at Oklahoma, who nurtured our creative growth. The other individual who gave me guidance early on was Sister Mary Martina, OSM, my high school journalism and chemistry teacher at Mount St. Mary's in Oklahoma City, who knew all along what path I should pursue.

Then, of course, I cannot forget my mother, Olga, and my late father, Henry R. They gave me encouragement, support, love, and a strong set of principles, not to mention a great sister, Jeanette, and two exceptional brothers, Johnny and Dennis.

Finally, the support and love from my wife Karen has helped me through not only this project but also some personal road bumps all along the way.

INTRODUCTION

In 1909, a courageous group of twenty-eight rubber chemists, members of the American Chemical Society (ACS), made a bold move and formed a subsection of the society—the India Rubber Chemistry Section—to discuss issues more pertinent to the industries for which they worked. While it took them another ten years to further define their identity as a division, the impetus for their almost rebel-like decision was not self-aggrandizement but a simple desire to share information about the properties of rubber. Foremost in their minds was the improvement of the quality of rubber coming from the wild jungles and plantations of the world's tropical zones. There was no consistency in quality or price, and they reasoned that *collective* minds working in harmony would help solve the problems or at least give them a better chance to deal with the issues. Furthermore, with a veil of secrecy draped over most industries at the junction of the nineteenth and twentieth centuries, there was no forum that offered them the opportunity to exchange data and thoughts without fear of reprisal.

The Rubber Division pioneers were right: Collective minds could indeed help their causes. One hundred years later, their success and the legacy they passed to those who followed can be seen in virtually every rubber or plastic product on the market today. From the tires that travel the world's roads to the vehicles that traverse outer space, rubber chemists and their professional organization, the Rubber Division of the ACS, continue to provide valuable information and support to a variety of industries in the interest of continuous improvement. Their mantra exists today: "To promote the education, professional growth and betterment of those individuals associated with the rubber and associated industries."

This book is not an attempt to reinterpret history, nor is it an attempt to enumerate the many events or mention all the individuals who were an integral part of the Rubber Division's past. A few publications already have gone into great technical detail about many of the elements noteworthy to the division's growth. In particular, *Rubber World*, originally called *India Rubber World*, has been documenting the division's actual development and activities for more than one hundred years. In relatively recent times, two particular issues were extremely valuable in the research of this book. Beginning with the October 1966 issue, four writers — Editor Otto Scott, Managing Editor Bill Mulligan, Technical Editor Joseph Del Gatto, and Engineering Editor Ken Allison — authored an excellent series of articles concluding in February 1967. The series was titled "The Division of Rubber Chemistry: Catalyst of an Industry." And in 1984, *Rubber World* published an equally impressive special issue commemorating the seventh-fifth anniversary of the division.

A competing publication to *Rubber World* was the now-defunct *Rubber Age*. Its longtime editor, Mel Lerner, not only provided solid editorial material about the division but also attended most biannual meetings and was one of the most active supporters of its efforts. Another publication that documented division activities in great detail is Crain Communications' *Rubber & Plastics News* (*RPN*). Founded in 1971 by Ernie Zielasko (who became familiar with the division when he was at *Rubber World* magazine), *RPN* also was an active division supporter. In 1984, *RPN* published a special about the division's seventy-fifth anniversary: *In Tribute to the Chemists Who Tame Rubber: Celebrating the 75th Anniversary of the ACS Rubber Division*. The following year, the division honored Zielasko with its Distinguished Service Award.

The final publication that recorded division events, *Rubber Chemistry and Technology* (*RC&T*), is the most complete. Founded in 1928, there is no periodical that covers the technical side of the division better than *RC&T*. The editors who solicited, collated, edited, and published thousands of technical papers contributed more to the industry's success than they realize.

The above-mentioned journals, as well as books on specific subjects (e.g., Maurice Morton's *Rubber Technology* and G. Stafford Whitby's *Synthetic Rubber*), are great contributions to any rubber chemist's library. This book cannot come close to replicating the knowledge and years of research devoted to such extremely worthwhile endeavors. Rather, it offers the nontechnical reader a morsel of information on rubber—its past, present, and future—and a condensed chronicle of some of the many individuals who made up the Rubber Division. After all, the division is comprised of people. My intent is to help the reader understand the lives of some of those people and provide an insight into the key events that drove the division and had an impact on the industry.

Great people often come from simple backgrounds, and it is in understanding this that we can better appreciate their individual and collective achievements. Through this book, as it reveals a little of their identities—often not at all related to rubber—the reader can come to know those members of the Rubber Division and their contributions to the industry.

The Rubber Division of the American Chemical Society did not conduct research into rubber or develop the compounds or the products. However, the division and its structure provided many rubber chemists with a collaborative "think tank" that fostered the ideas to help many individuals develop those creations so vital to our lives today. In a December 10, 2007, edition of *Rubber & Plastics News*, Managing Editor Bruce Meyer wrote of Daniel L. Hertz Jr., the seventy-seven-year-old president of Seals Eastern, Inc., "Hertz is a self-educated scientist who got much of his education through ACS Rubber Division courses, attending conferences and reading the hundreds of science and technology books in the library outside his office." Hertz was selected that year as the publication's Rubber Industry Executive of the Year.

But longtime division historian Ben Kastein said it best. Recalling the many technical papers contributed by division members, he said, "The division is all about relationships, information. The papers reflected the current interest at that time. If you look at the papers that had been published, that gives you a trend of what was important in the rubber industry. So the Rubber Division was kind of a mirror of what was (and is) going on in the rubber industry."

PART ONE

....

IN THE BEGINNING

....

KING
OF THE
JUNGLE

1

> *The most beautiful thing we can experience is the mysterious. It is the source of all true art and science.*
>
> ALBERT EINSTEIN

In the jungles that cover the earth's tropical zones like broccoli heaped on a salad, there are trees that ooze a milky substance when lanced by a spear, a rock, a knife, or any other sharp object. Indigenous people of those areas were the first to notice that the liquid would gradually harden and turn black on exposure to air.

As near as anyone can determine, the French were the first civilized group to conduct a study, of sorts, on this unique material. In 1735, under the leadership of mathematician Charles-Marie de la Condamine, surveyor Pierre

Bouguer, and scientist Louis Godin, the Académie Royale des Sciences in Paris dispatched an expedition to Peru, ostensibly to calculate the size of a degree of latitude at the equator. They found that the country contained something more than mountains and people.

Strangely, the three men finished the journey to their ultimate destination by different routes, Bouguer and Godin sailing to Quito and de la Condamine traveling overland from Manta. The three converged at Quito and were soon arguing, something of a modus operandi for scientists. Godin wanted to work on his own, while Bouguer was content to work with de la Condamine. Bouguer proceeded to measure the density of the earth using the instruments of his time—a plumb line and a mountain. He and de la Condamine eventually talked to each other, and the two later collaborated on a published account of their work, *La Figure de la terre* (1749).[1]

The odd triumvirate completed their measurements in 1743, and to none of the three's surprise, they returned to France by different routes. Before he left Peru, de la Condamine had found something else in his expedition that he wanted to share with colleagues back at the academy in Paris. Intrigued with some substance the natives were manipulating, he had gathered a sample from them and sent it back to the academy with a note titled "Caoutchouc:"[2]

> There grows in the province of Esmeraldas, a tree called by the natives of the country "Heve." There flows from it, by simple incision, a [liquid], which hardens gradually and blackens in the air. The inhabitants make flambeaux of it, which burns very well without wicks, and gives rather of a fine light. . . . In the province of Quito, sheets of linen are coated with it, and are used for the same purpose as we use wax cloth. . . .
>
> The same tree grows along the banks of the Amazon and the Mainas (Mayan) Indians call the resin which they extract from it "cahuchu." They make boots of it, which do not draw water, which after being blackened by being held in the smoke, have all the appearance of leather. They coat earthen molds in the shape of a bottle with it, and when the resin has hardened, they break the mold and force out the pieces through the neck and mouth: thus they get a non-fragile bottle, capable of containing all kinds of liquid.

The Peruvians, the other South Americans, and their inquisitive European visitors were dealing with what we call today natural rubber. They named it "caoutchouc."

Before de la Condamine departed, and likely tired of measuring mountains, he made a side trip to Cayenne, the capital of French Guyana in northeast South America, where he met fellow Frenchman François Fresneau, with whom he shared his experiences, including the material the Peruvian natives were manipulating. Fresneau became intrigued by de la Condamine's stories and, in particular, by the new substance, caoutchouc, that he was shown. About ten years later, Fresneau gave de la Condamine a report to present to the Académie Royale. This was the first methodical paper on rubber, published in 1755. De la Condamine used Fresneau's report as a basis for a paper he delivered to the academy titled "Memoir on an Elastic Resin Newly Discovered in Cayenne by Monsieur Fresneau, and on the Use of Various Milky Juices of Trees of Guiana of Equinoctial France."[3] In spite of its lengthy title, the report is one of the first known documents by a European who considered the substance as a potential industrial material.

De la Condamine wasn't as interested in the new substance as Fresneau, who spent the next ten years studying rubber, using exploration and scrutiny as his research tools, documenting his observations along the way. His later-published papers describing his findings didn't make any strong scientific claims, but Fresneau at least saw the budding possibilities of rubber after studying its properties for so long.

While the French are believed to be the first to study caoutchouc, or rubber, their neighbors to the south, the Spaniards and Portuguese, were aware of the substance much sooner. Some historians note that about 1519, a Spanish writer, Antonio de Herrera-Tordesillas, reported on Aztec children playing games with balls made "from a black resin, obtained from a tree called by the natives 'ulaquhuil.' " The name seems to have been a derivation of a tree, the *Castilla elastica*, or *ule*, which still grows in parts of Mexico and Central America. Other chroniclers of events point out that Pierre Martyr d'Anghiera, another Spanish writer from Madrid on a visit to Mexico in 1525, also published a description of some children playing with balls made of that substance.

Normally, given the era and its absence of today's sophisticated communications tools, media, and so on, either of the authors' writings might be interpreted as purely anecdotal. However, the credibility curve is enhanced by knowledge of Anghiera's background. He was employed as a tutor to the children of King Ferdinand II of Aragón and Queen Isabella of Spain and became prominent as a writer with two major works: *Opus Epistularum* and *De Orbe Nouo*. Most pertinent to his documentations about rubber balls in Mexico, *De Orbe Nouo* is a work that has extremely important historical value in U.S. history, as it documents Christopher Columbus's discovery of America. Anghiera had access to more than 816 letters addressed from Columbus to the royal court, as well as many of the individuals who made the journey. It is unknown if he actually talked directly to Columbus.

On his second voyage to the New World, Columbus also reported some natives of Haiti playing with a ball made from the sap of a tree. Regardless of who—or what country—first mentioned "rubber," no one will dispute the expeditions by the French, Spanish, and Portuguese pioneers that brought about the substance's introduction to the rest of the world. However, for almost three hundred years after Columbus, there was still no commercial use for natural rubber and it was, for the most part, just like the surroundings of its source—very primitive, uncontrollable, and not travel-friendly. Other than rubber balls and other items quickly molded in the jungle, no one saw the potential for the substance. It would thicken fairly rapidly in the air, making it virtually impossible to transform it into anything more marketable and usable than those rubber balls.

In the latter part of the eighteenth century, however, the English got into the game big time. On April 15, 1770, British philosopher and self-taught chemist Joseph Priestley, who would later be recognized for his discovery of oxygen, and other compounds, discovered that caoutchouc rubbed out lead pencil marks. In one swift moment, he invented the eraser and then gave the material the name "rubber." About that same time, Priestley's new name acquired a one-word prefix, "India." In early English trading circles of that era, caoutchouc was classified as an "exotic." Exotics were likely the name given to what we today call commodities. As such, all exotics were directed to a section in the southeast corner of London, near Seething, Mark, and Mincing lanes, that was the port of entry for ships of the East India Company, the largest commercial venture in the world at that time. Any savvy salesman who wanted to sell exotics knew he had to deal with the East India merchants. So, probably in the cockney sort of way, to give a quick description of the product to potential buyers, and because it came from the East India Company (not India), it became known as India rubber.[4]

The product now had a new name, but to reach the point of practical usage—a viable application—early rubber chemists and researchers of the day knew they had to spend the majority of their time trying to find a good solvent for this new substance, something they could transform into a workable substance.[5] As de la Condamine and others later discovered, as soon as you take it out of the jungle, it coagulates quickly. The first solvents tried were turpentine and ether. Two Frenchmen, a chemist, Pierre Joseph Macquer, and a physician, Jean Thomas Herissant, tried turpentine, but soon discovered that their efforts were greatly affected by temperature; in hot weather, the rubber became sticky, and in cold weather it became hard and stiff. In 1773, Macquer tried to make some riding boots for Frederick the Great by coating a wax cylinder with rubber dissolved in ether. The ether evaporated quickly, and the wax, of course, melted in hot water.

In 1791, another French chemist, Antoine François de Fourcroy, discovered that "fixed and volatile oils" turned rubber into a "varnish" that could be easily spread on fabrics.[6] A year later, the first rubber patent was issued to Samuel Peal of England, using turpentine as a solvent to waterproof fabrics. In 1818, a Scottish surgeon, James Symes, distilled coal tar to get a straw-colored liquid that provided a better solvent—essentially naphtha. It was cheaper than turpentine and enabled the rubber to dry better. Five years later, a patent was issued to another Scot, Charles Macintosh of Glasgow, for a waterproof fabric of two pieces of wool held together with rubber using Symes's naphtha solvent. Its application? A coat to repel rain. Today, the British still call the raincoat a "macintosh" or "mac."

Initially, Macintosh's product wasn't as successful as he wanted it to be, but a London coach maker, Thomas Hancock, thought Macintosh's product worthy enough to use on the top of his carriages and acceptable for other things, such as curtains, pant-pocket edges, shoe tops, and so on. Hancock became so enamored—and successful—with his new diversion that he quit the coach business and went into producing more of the new rubber material. He later would manufacture some solid rubber covers to serve as cushions for the wheels of Queen Victoria's carriage and be credited, in many historical circles, as the catalyst of today's tire industry.

For the next two decades rubber chemical discoveries took an unintentional sabbatical—until Charles A. Goodyear's revelation of vulcanization in 1839. But even then, Goodyear didn't reap any monetary benefits from his discovery because Hancock had beaten him to the patent punch by securing rights in England to a sulfur-heat rubber treatment. Goodyear eventually received a U.S. patent on June 15, 1844, but by that time, natural rubber already had developed a significant amount of attention throughout the world.

With Macintosh's application and the influx of rubber shoes from Brazil, natural rubber had a couple of its first marketable products. But even by the middle of the nineteenth century, demand for the product hadn't yet forced the jungle workers—primarily in the lower Amazon region of Brazil—into overtime. There were a few reasons this substance took its time in developing. For one, the process of gathering the material—essentially from the tropical forests in South America's Amazon basin—and transporting it by canoes to the ships was very time-consuming. Then the voyage across the ocean was equally lengthy. The other chink in the process was in the trees themselves. From seed to tapping, it took seven years for a tree to mature. In spite of those relatively small hurdles, the demand for natural rubber started to increase about 1860 primarily because of a civil war developing in the United States. That sent the price of natural rubber up to more than one dollar a pound in 1865.[7] It wouldn't be the last price squeeze, nor would it be the last regional or world conflict that would have a significant impact on rubber's supply and demand.

In spite of the increase in demand, the supply of Brazilian rubber, *Hevea brasiliensis,* continued to grow and it precipitated the dividing of the Maranhão and (Belém do) Pará regions and the formation of Amazonas where a higher quality of hevea existed. The problem for the Amazonas, though, was that it was several hundred miles up-river from Belém, and overtapping and slashing of trees to meet the increase in demand would have an obvious and negative long-term impact.

In 1866, one leader, Don Pedro II of Brazil, thought he could mitigate some of the country's problems. He had the foresight to capitalize on Brazil's natural resource and opened the Amazon River to ocean-going steamships that enabled those vessels to travel as far as fifteen hundred miles inland to load their products. The likely unfortunate aftermath of this decision, however, was the increase in production that forced more inexperienced tappers (called *seringueiros*, or "men of the syringe tree") into the harvesting process, which killed thousands of trees because of improper tapping.

In spite of his efforts, Don Pedro's prudence was short-lived because the problems experienced in the jungles of South America became opportunities to those outside of that continent. One individual prominent in the development of other geographical sources of natural rubber was Sir Clements Markham, a former officer in the Royal Navy turned geographer, entrepreneur, and opportunist. Through Markham's prior success in transplanting cinchona (quinine) seeds from Peru to India, he knew he could do the same with hevea seeds from Brazil.

In 1873, through James Collins, curator of Physic Gardens of London Apothecaries Company, Markham enlisted the assistance of Brazilian-based Henry Wickham to procure the seeds. Wickham hired some natives to collect, pack (between layers of banana leaves in wicker baskets), and hang the seeds in the hold of a ship to allow air to circulate and prevent them from turning rancid. On June 14, 1876, the ship docked at Liverpool, England, and the seeds were transported by rail to the Royal Botanical Gardens at Kew, outside London.[8] By 1880, the seeds that had been successfully planted in Kew were sprouting and thriving in small experimental plots on estates in southern India, Ceylon, Singapore, and Java. Later, that list would include Central Africa. *Hevea brasiliensis*, natural rubber, had now truly become global.

PART TWO

PARENT

GENESIS 2

*Tell me, and I'll forget. Show me, and I may
not remember. Involve me, and I'll understand.*
NATIVE AMERICAN PROVERB

In August 1874, in the small east-central Pennsylvania town
of Northumberland, a group of chemists gathered for what
may have been a defiant meeting at the time. What was
their rebellious idea? They wanted to celebrate, on the
centennial of the historic event, the discovery of oxygen by
one of their own—Joseph Priestley. Priestley was one of
their idols. Northumberland also happened to be the final
resting place of the British philosopher-turned-scientist
and Northumberland's adopted son. The meeting evolved
at the initial suggestion of Henry Carrington Bolton of
Columbia College.

Bolton had written a letter that was published in the April 1874 edition of the *American Chemist*. Noting the date and the coincidence of the one-hundred-year anniversary of Priestley's discovery (and the discoveries of other notables, such as Carl Wilhelm Scheele, Antoine Laurent Lavoisier, Charles L. Cadet de Gassicourt, and Karl Georg Lucas Christian Bergmann), Bolton wrote that 1774 might well be regarded as the birth year of modern chemistry and suggested that it "would be an agreeable event if American chemists should meet on the first day of August 1874, at some pleasant watering-place, to discuss chemical questions, and especially the wonderfully rapid progress of chemical science in the past one hundred years."[1]

Editors of the *American Chemist* thought the suggestion had some potential, likely from a marketing, advertising, or PR perspective. So they quickly invited chemists around the country to write to them and express their opinions about Bolton's proposal. The idea kindled particular interest from a young professor of chemistry at the Woman's Medical College of Pennsylvania,[2] Miss Rachel L. Bodley. "I made a pilgrimage last August to the grave of Priestley in Northumberland, Pennsylvania," the professor wrote, "and was deeply impressed by the locality, its associations and its charming surroundings. My proposition is, therefore, that the centennial gathering be around this grave, and that the meetings, other than the open air one on the cemetery hilltop, be in the quaint little church built by Priestley, where might be exhibited the apparatus devised by the great scientist and used in his memorable experiments."

Bodley's suggestion was resplendent in several ways. The editors agreed that her suggestion was refreshing and very plausible, and they quickly contacted Bolton, who, on May 11, 1874, called a meeting of his local professional group, the Chemical Section of the New York Lyceum of Natural History. Its members adopted a resolution that parted the suggestive seas and allowed reality to cross their minds. It read in part: "A committee of five [will be] appointed by the Chair, whose duty it shall be to correspond with the chemists of the country with a view to securing the observance of a centennial anniversary of chemistry during the year 1874."[3] Bodley's suggestion not only made her popular among a few chemists in the country but also gave her a more-than-casual position in the history of chemical organizations in the United States.[4] The five-man committee was appointed by Bolton, and the committee members made all the necessary preparations and invited chemists of the country to assemble at Northumberland on July 31, 1874.

Northumberland is located in Pennsylvania's quaint Susquehanna Valley. Woodlands cover about half of the land area of the county, but much of the nonmountainous areas are rich and fertile farmland. Today, anthracite coal mining, quarrying and other heavy and light manufacturing and distribution operations are the county's major enterprises.

About eighty chemists made the pilgrimage to Northumberland and attended the celebration, the first session of which was held in the public school building. There were speeches delivered and congratulatory cablegrams exchanged, an unveiling of a statue of Priestley, readings of Priestley's letters, and a memorial address delivered at Priestley's grave by Henry Coppee, president of Lehigh University. The chemists were there ostensibly to honor Priestley and other notable chemical science forefathers. Also on their agenda, however, was a discussion about promoting their avocations and becoming more organized in their efforts to spread the gospel of chemistry.

A perfunctory attempt at unification for a common cause had been made the year before, in 1873, at an informal gathering of chemists who were attending the regular meeting of the American Association for the Advancement of Science (AAAS) at Portland, Maine. With the support of Professor Samuel A. Lattimore of the University of Rochester, a group of chemists agreed to ask the governing body for permission to form a subsection of the organization for the purpose of studying chemistry. Permission was accorded, but the subsection didn't officially become a section (C) until 1881.[5]

In the meantime, the 1873 attempt notwithstanding, the Northumberland gathering pushed forward. On the afternoon session of July 31, 1874, several of the AAAS-affiliated chemists at the Northumberland meeting, still concerned about autonomy, believed that more "freedom of action could be secured by forming, in addition, an independent national society."[6] So at the suggestion of Persifor Frazer, a professor of chemistry at the University of Pennsylvania and a Civil War veteran of the Battle of Gettysburg, the group was urged to form a "chemical society which should date its origin from this centennial celebration, and since America had not a single society to represent the chemical thought of the country."[7]

With expected opposition from J. Lawrence Smith, a recent past president of the AAAS, the group agreed to amend Frazer's motion and lobby for subsection status in the AAAS. On August 17, 1874, at the society's Hartford, Connecticut, meeting, this status was achieved. Seven years later, in 1881, the group would finally receive status as a separate section (C), which still exists today. However, many of the dissidents who were at the Northumberland meeting weren't fully satisfied with this "1874 compromise" and still wanted a society of their own. On January 27, 1876, in the New York City home of Charles F. Chandler of the Columbia School of Mines (University), a meeting of chemists with the intent of forming their own organization was held and a committee formed, consisting of Messrs. Chandler, Habirshaw, Endemann, Alsberg, Morton, Walz, Hoffmann, and Casamajor, to pursue this formal organization.

Finally, on April 6, 1876, at the College of Pharmacy in New York, the new society was officially organized and, two weeks later, its members elected sixty-five-year-old New York University professor John William Draper as its first president. William M. Habirshaw, a former U.S. Navy officer and well known throughout the rubber trade of that time, was elected treasurer. The first regular meeting after organization was held on May 4, 1876, in the University Building at present-day New York University, with thirty members and fifteen visitors present. More than a year later, on November 9, 1877, the American Chemical Society was legally incorporated under the laws of the State of New York. Years later, in 1908, the membership, likely cognizant of the organization's struggle for autonomy and perceptive regarding other members' interests in other chemical pursuits, formed its first five divisions: Industrial Chemists and Chemical Engineers, Agricultural and Food Chemistry, Fertilizer Chemistry, Organic Chemistry, and Physical and Inorganic Chemistry.

In 1875, while the American Chemical Society was in its formative stages, a young man from Vermont arrived on the campus of Buchtel College in Akron, Ohio. Charles Mellen Knight was too young to enlist in the Union army, so he went westward to Ohio to assist college students in the instruction of science. George W. Knepper, in his book *New Lamps for Old*,[8] a comprehensive history of the University of Akron, wrote of Knight, "In those days when chemistry was taught more as a cultural subject than as a narrowly professional one, Knight was best remembered for his ability to show how chemistry related to daily life and activity. Often he employed illustrations derived from his own research and from his industrial consulting. He was as much admired for his thoughtful concern for his students, even long after they had graduated, as he was for the clarity of explanation that did so much to lead some to — and others through — the complexities of science."

By the time he retired in 1913 and passed his teaching duties to Hezzelton E. Simmons, Knight had laid a foundation of chemistry and chemical education unequaled in any generation and established a solid base for what would evolve into a sophisticated and influential field of endeavor that would impact the world — rubber chemistry.

PART THREE

YOUTH

LEAP OF
FAITH

3

You must be the change you wish to see in the world.

MAHATMA GANDHI

At the corner of the nineteenth and twentieth centuries, the world was in technological puberty and waiting for the run-don't-walk sign. In the United States, from 1900 to 1910, the industrial age was cranking like a war might break out, and mass production pushed prices to all-time lows. It was the first decade of a strong focus on manufactured goods. The consumer was king and queen, and while the Bible was still the best-read book, the Sears & Roebuck and Montgomery Ward catalogues were running a close second.

Even the president of the United States, Theodore "Teddy" Roosevelt, was squeezed into the commercialism incubator. A newspaper cartoon started a fad of "Teddy Bears," which hit the factory lines in 1905. Smokestacks and their telltale, black coal-burning clouds could be seen on ships at sea, on trains overland, and from the big city factories that protruded from the horizon like huge burning pencils. Radios blared music, commentaries, and a variety of dramatic programming, and the world began taking off in many new directions. None, however, was quite as mesmerizing as the new, gasoline-powered vehicle that didn't need a track.

The device that would change the world was the automobile, and just as important, the residual effect of Henry Ford's development of this new machine would, over time, jump-start thousands of ancillary industries. One of those industries already was both literally and figuratively a bouncing baby, and the child was getting ready to challenge the world but needed a little guidance. Of course, that child was the rubber industry, and while rubber may not have been the king of commerce at that time, it was certainly on the king's court.

By 1900, the rubber industry had taken a sticky grip on the world. The universal adoption and improvements of Charles Goodyear's vulcanization process since 1850 and the increased use of electricity and steam as sources of power provided a growing market for anything made of rubber.[1] While those would-be kings were cars from Ford and other entrepreneurs, it didn't take a Nobel Prize winner to understand that if their innovations were successful, any industry that supplied products to an automobile venture also would reap the benefits of the alliance. In 1903, Ford's company alone built more than 1,700 cars, and by the end of 1910, that figure had risen to more than 32,000 vehicles. As an industry, in 1900, there were 8,000 automobiles registered in the United States. In ten years, that figure exceeded 468,000, and in 1913, registrations topped the 1 million mark.[2] Because of the new phenomenon they called the automobile, the rubber industry appropriately was positioning itself to take advantage of the potential opportunities presented by the success of Ford and his colleagues.

Rubber not only was gaining in popularity but also growing in the variety of its applications. Even with the scarcity of wild rubber in the

Brazilian jungles, there were more than three hundred different grades of rubber and rubber-like gums reported; and there were as many as twenty grades of Brazilian rubber, twenty other South American grades, ten Central American grades, and fifty African grades on the market. By 1914, plantation rubber exceeded the amount of wild grades on the market, and from that time on, the number of varieties marketed declined.[3]

Most of the rubber wasn't used for balls, bicycles, or horseshoe cushions anymore, and the rubber brokers knew prices certainly could only get better for them. The price for natural rubber was rising, reaching three dollars a pound in 1910, primarily because of demand from the auto industry and its tires—and it would multiply.[4] The majority of that premium rubber at the start of the decade was going into tires, referred to as "casings" in the early stages of record keeping, and they accounted for 2.4 million units. Including tubes, that figure increased from 2 million to 4.4 million.[5]

In his article "Bird's Eye View of Rubber in North America" (August 1909), *India Rubber World* editor Henry Pearson recorded that the 90 million pounds of rubber the industry produced annually yielded 600 million pounds of finished goods, containing 200 million pounds of metallic oxides and clays and 150 million pounds of fabrics. The most common rubber products at that time were belting, hose, packings, and footwear. Tires still were not at the top of the list but were the next category, and because of the embryonic auto industry, the potential for rubber tires was massive.

Rubber had progressed from its horticultural roots of *Hevea brasiliensis* to thirteen varieties of Brazilian Pará rubber, including islands, upriver, caucho, fine, coarse, new, and old; twelve varieties from Africa; eight from Central America; three from East India; and one from Ceylon—a minimum total of thirty-six varieties and types.[6] With the popularity and increased usage, however, came other concerns. The many varieties, multiple sources, and even variety of prices were the auxiliary issues that now confronted the industry. While mass production had yet to impact the manufacturing process—and it would take another ten to twenty years for that to occur—chemists were the only ones who could possibly deal with these variables during processing and help find the lowest-cost alternatives for their companies.[7] The chemists were charged with turning raw rubber, whatever the type or source, into black gold.

However, while the chemists could perform their magic with rubber recipes, they couldn't manufacture the products. They needed help. Pearson also reported there were 260 factories producing goods, employing fifty thousand workers (or one hundred thousand if you included the plants supplying textiles, machinery, and other materials). But before the rubber industry could take advantage of the potential business in the auto market, they needed to become more cost-efficient with their production efforts. That's where the chemists could help.

To begin with, the chemists had to help themselves with pricing issues. The industry had to find a lower-cost source of natural rubber. In 1908, upriver Pará, one of the top grades of natural rubber, had boomed to a high of $1.30 a pound from a low of 79 cents, but ironically, this inflation in the price of the top grades of rubber made the lower grades, such as Africa-grown Pontianak, more appealing, and that helped the sourcing search for a while. Pontianak was selling for 5 cents a pound, and even guayule (an alternate source of natural rubber that comes from a shrub native to northern Mexico and the southwestern part of the United States) was selling for 30 cents a pound, but Pontianak and the other alternatives, including reclaimed rubber, couldn't hold a beaker to Pará in quality and consistency of product. Quality was always an issue, and it typically was different with each variety.

In the meantime, Brazilian rubber, especially the premium Pará, was becoming increasingly more difficult to acquire, due to both the rise in price and the decrease in availability. Export taxes were excessive compared with the smaller taxes imposed on Asian rubber. In addition, the infrastructure in jungle-producing areas of Brazil—waterways or rivers—needed frequent dredging to allow ships access to the ports, an expensive and time-consuming process. To add to the dilemma, those ports were in great need of continuous maintenance and upgrade due to the increase in usage.[8] So even if you could get Pará rubber, it was going to cost even more than its usual premium price.

Concurrently with the problems in Brazil, shipments to the United States from Africa and Asia began to increase, and while the boom was more like a very slow crescendo, wise shoppers were now looking to the

East for their raw materials. While natural rubber prices were dropping, so too was the quality of the product. The fledgling rubber industry was caught in a somewhat opaque situation. Mainly because of the growth of the auto industry and the subsequent need for tires and other automotive rubber products, now the pressure shifted to the chemists to find "magic formulas" that would help solve those issues dealing primarily with quality, estimated to represent 30 to 40 percent of the total cost of a tire.[9] In short, the demand for rubber continued to escalate, but there was no consistency in the price or in product quality, nor was there a solid, universal means to measure that quality. Furthermore, there was no unified or collaborative effort on the horizon that would assist the industry in addressing those issues on an industry-wide basis. Without any organized effort to conduct any kind of joint problem solving, the next logical step was to look for help from the greatest users of rubber — the tire companies.

While the United States led the world in the export of natural minerals such as copper, zinc, coal, and petroleum, among others, U.S. tire makers had lagged behind the French during the bicycle craze of the 1890s and only gained an advantage in productivity when the automobile industry cranked up its assembly lines right before World War I.[10] It was still too much to expect that the industry might be able to control the quality of the natural rubber coming from the wild and some of the newer plantations. However, it was not beyond comprehension for the manufacturers, especially the tire companies, to establish quality controls within their own industries to ensure consistency for their customers. With the demand for rubber burgeoning, manufacturers had to develop, among other things, reliable testing methods for their raw materials, and the methods had to be usable and acceptable throughout the industry. Some companies developed improvements in rubber chemistry, but they were either accidental or fashioned by individuals in the factories who were untrained and lacked any theoretical knowledge of the subject and who depended on the trial-and-error method of using a random range of additives.

If there were pockets of success in those attempts, the methods were often figuratively and literally hermetically sealed in guarded rooms. Information the companies and their chemists gathered was top secret and was not shared. Few chemical innovations were patented at that time. If someone did share knowledge with a competitor—even in a casual or social vein not intending any covert activity—that person faced being fired and blacklisted throughout the industry. It was a very secretive era filled with plenty of paranoia.

In the meantime, there was a group of people who had the foresight, the impetus, and the esprit de corps to approach the rubber industry's concerns without any desire for fame, fortune, or reciprocity. They were frustrated with the curtain of proprietary secrecy draped in front of the great hall of rubber chemistry. Their discontent would partially end on the second to last day of the year in 1909. At a meeting in Boston of the American Chemical Society, twenty-eight chemists with unyielding convictions toward a unified approach to address issues of sharing information for the common good—specifically of the rubber industry—formed the India Rubber Chemistry Section of the American Chemical Society.

Nevertheless, there were still obstacles to the objectives of these dedicated men. The unselfish devotion these members had to a unified cause and to one another were tantamount to rebellion in those days. Cognizant of how and why their own American Chemical Society had been formed, the men had to overcome historical precedent and tradition, to challenge the status quo of one of the nation's premier and largest professional societies. In addition, their action was coming at a time when funds for research were not readily available to chemists. Frankly, at that time, the rubber industry didn't really know what a chemist was or what he could do. These men were taking a classic leap of faith.

Charles Cross Goodrich, the eldest son of Benjamin F. Goodrich, founder of the B. F. Goodrich Company, was given the responsibility of leading the new India Rubber Section into the semicharted jungles of the industry woes when he was elected chairman of the section. Charles graduated from Harvard in 1893 and attended the Massachusetts Institute of Technology and Lawrence Scientific School at Harvard for postgraduate work in science. In 1895, after working alone in a small laboratory on Factory Street in Akron, his father hired him and Charles established the first chemical laboratory, believed to be the first in the industry.[11]

Frederick J. Maywald was named secretary of the section, and other pioneer members included William Geer, a researcher from B. F.

Goodrich who would later invent the Geer oven test; George Oenslager, a chemist from Diamond Rubber who discovered the use of organic accelerators, among other achievements; C. R. Boggs of Simplex Wire and Cable, who would later serve as chairman of the section; Edward A. Barrier; Milton E. MacDonald; William G. Hills; Sheldon P. Thatcher; Harvey M. Eddy; H. Hughes; Harold van der Linde; M. L. Allard; C. E. Waters; and G. H. Savage. The section was organized as a collator of information related to the field of rubber chemistry. Its function, as described much later on by *Rubber World* editor Ed V. Osberg, was "to bring together at occasional meetings, chemists and others interested in the field of rubber so that problems of moment and interest could be solved by common effort."

Like a parent looking apprehensively at a child, initially, the India Rubber Section was not met with universal approval. In September 1911, *India Rubber World* published the following statement issued by the American Chemical Society:

> The American Chemical Society is very anxious that those of its members interested in the chemistry of India rubber should have their problems considered and solved. The India Rubber Section has had two meetings, but there is not yet sufficient evidence of real cooperative effort among the rubber chemists to insure success. The methods of analysis of India rubber are in almost a chaotic condition. The usual specifications for rubber goods meet the approval only of those who make them. The general chemistry of India rubber is sadly in need of improvement.
>
> Only the chemists actively interested in the India rubber industry can hope to improve affairs and it is, accordingly, necessary that they should really get together without too many padlocks on their lips if results are to be accomplished. It is certainly true that there are secrets of the rubber trade which cannot be disclosed, nor is there any desire that they should be disclosed, but when certain firms decline even to allow their methods of analysis to be known, it would certainly seem that secrecy is carried too far. The Section can never become a success if every member goes to its meetings with no idea of responsibilities toward helpfulness, but simply to learn from others, many of whom may be in a similar position.

India Rubber World was the first magazine devoted to coverage of the industry. Established in 1889, the initial goal of the publication was to "aid materially the scientific and the mechanical development of business in India rubber gum, gutta-percha, and kindred products, by giving the manufacturer all meritorious information procurable as to old and new methods and compounds . . . knowledge . . . as to the workings of factories . . . the results of practical experimentation according to satisfactory formulas and desirable patented processes . . . new inventions . . . patents granted . . . and new goods introduced to the market," wrote P. F. Mottelay, the magazine's first editor. However, after just three issues, Mottelay sold his interest in the paper to Henry Pearson.[12]

The members of the India Rubber Section might have taken the ACS's published statement as a disapproval of their actions; instead, they accepted it as constructive criticism and as a guideline, of sorts, for improvement—a challenge.

Predictably, the traditional pattern established by the American Chemical Society was working against them, yet the group was buoyed by earlier successes from two of their own. In 1899, Arthur H. Marks invented an alkali reclaiming process that would serve as a significant milestone in the development of the reclaim rubber industry. This process would serve the tire and rubber industry well into the twentieth century. In 1905, Marks left his position at Harvard University to join the Diamond Rubber Company (purchased in 1912 by the B. F. Goodrich Company) in Akron, Ohio. He brought with him a colleague, George Oenslager. In 1906, Oenslager provided the ACS's other benchmark achievement when he discovered organic accelerators. Oenslager's timely findings, a derivative of aniline called thiocarbanilide, could accelerate the action of sulfur on rubber. This meant that rubber could cure in much shorter time, ultimately saving time and money. Organic accelerators have since become a staple in the industry.

Benjamin Kastein Jr., division historian for fourteen years from 1978 to 1992, recalled a story about Oenslager and an assignment that Marks had given him. "Oenslager was to find a magic ingredient that would improve the properties of Pontianak rubber," said Kastein.[13] "In a short time of investigation he came up with a catalyst. He added a few ounces of aniline added to the compound and it accelerated the cure. . . . Normally a passenger tire would take ninety minutes to cure, but with the aniline compound, it would cure in thirty minutes. So you can imagine if you were a production manager, and all of a sudden you got three times the production out of the same equipment. That would be a wonderful invention."

The problem with aniline, though, was its toxicity. In early manufacturing, the mixing was all handled in one building, but the aniline would vaporize and drift over and settle on white rubber and turn it yellow, so the plant built a tent outside for the aniline mixing. But management was warned to keep a close eye on the workers, "and when [an employee] began to stagger," said Kastein, "they would grab him by his shirt and take him as fast as they could outside and set them on a bench to recover, then feed him a quart of milk . . . the antidote . . . then send him back into the tent again to finish the mixing."

In 1912, Diamond changed the mixture to a less-toxic product that worked even better, cutting curing time to about ten minutes. On a later trip to England, Oenslager also discovered the use of carbon black as a reinforcing filler instead of zinc oxide, commonly used in the United States. In the 1940s, the oil-furnace process of carbon black manufacture was introduced, which allowed plants outside the United States to produce their own carbon black and save exporting costs. This would prove its importance to the World War II synthetic rubber effort because the new GR-S required considerably more carbon black than natural rubber, which in many applications required very little.

Some historians have said that Oenslager's discovery of organic accelerators was the next most significant achievement after Goodyear's invention of vulcanization. Irrespective of the achievements of Marks and Oenslager, the small group of pioneers who established the India Rubber Section were well on their way to making a solid impact on the rubber industry.

In the meantime, the progress of Oenslager's India Rubber Section wasn't moving at the pace he exhibited in his own discoveries. The section was moving as slowly as latex down the cut bark of a rubber tree. The group didn't have another formal meeting for another two years. In 1911, they had two meetings—one in Indianapolis in June at which only sixteen men showed up, and the other in Washington, D.C., in December, at which an improved number of fifty-four attended. In April 1912, only thirty-six members attended, and most hesitated to talk about anything other than the weather.

Sensing an urgent need for a timely objective, Chairman C. C. Goodrich moved quickly to help focus the section's mission and the industry's major concerns: How to improve the quality of rubber coming from several sources, how to measure that quality, and how to develop standardized testing methods. With those in mind, they formed committees whose responsibility it was to address some of the issues. One of those first committees was chaired by Charles M. Knight, and that group was organized to develop standards on rubber testing—chemical and physical—of manufactured and crude rubber. Knight's selection and area of expertise was appropriate since he had just established the first courses in rubber chemistry at Buchtel College in Akron, where he was a professor of chemistry. They decided to start first with compounded rubber.[14]

Charles Knight was accustomed to volunteerism and righteous causes. It was a family thing. He and his twin sister, Helen, were the youngest of eight children born to farmer and sometime-teacher Joel Knight Jr. and Fanny Maria Duncan. The Knights were a forthright and religious family. The oldest child, Sophie, married Sullivan McCollester who was a Universalist minister and pastor of a church in Nashua, New Hampshire. In 1872, Henry Blandy, a trustee at a newly formed college in Ohio organized by the Universalist Church, summoned Sullivan to Akron to gauge his interest in becoming president of the new college. McCollester accepted the board's offer and assumed the title of president of the faculty of Buchtel College.[15] McCollester held that position until 1878, when he resigned.

In the meantime, Charles Knight graduated from Westbrook Seminary in Deering, Maine, and then enrolled in Tufts College in Medford, Massachusetts. Charles's continuing education at Tufts was temporarily suspended when he left school for three years during 1869–70 to help his brother-in-law, former Civil War colonel Henry "Hal" Greenwood, in the construction of the Kansas-Pacific Railroad, working primarily as a bookkeeper in the engineer's office. After his work was completed,

Knight returned to Tufts in 1873 and graduated Phi Beta Kappa. In 1875, he went to Akron and Buchtel College.

Knight served Buchtel College from 1875 to 1884 as professor of natural sciences. In 1884, he was appointed professor of chemistry and physics and became interim president of Buchtel College during the 1896–97 school year. In 1897, he was given an honorary doctor of science degree and appointed the first dean of the faculty. In 1908, he organized the college's first class in rubber chemistry—the first of its kind in the country. He continued to teach chemistry and serve as dean until 1913, when, after thirty-eight years of teaching, he retired. He is also credited with initiating the university's varsity basketball program.

In the meantime, Knight's help with his committee on rubber testing for the India Rubber Section spawned another focus group in 1911, the Joint Rubber Insulation Committee, a unique convergence of manufacturers and consumers of insulated wire that supplied a great deal of data on tests used by the India Rubber Section. In a December meeting of that year, the section increased its focus on issues of the day. Section chairman D. A. Cutler first introduced the topic of synthetic rubber and experiments with the product from isoprene obtained from turpentine. Then F. E. Barrows made a short presentation on the formation of the rubber molecule. Indeed, the section was beginning to increasingly reflect the industry's realities of the day.

Many years later, in May 1990, longtime division historian Kastein presented a paper at a Rubber Division meeting in Las Vegas titled "People Make the Difference." The source of his information was a series of taped interviews of key members and Rubber Division participants:[16]

The Las Vegas meeting of the Rubber Division, ACS, provided attendees the opportunity to hear the interview of Mr. Arnold H. Smith, by Mr. Herbert A. Endres, recorded April 7, 1966. Mr. Smith, as Secretary-Treasurer of the Division from 1919 to 1928, and as Chairman in 1929, was the person most responsible for laying the foundation, which supported the growth of the Division to its present status.

. . . Smith heads my list of "People Make the Difference." Significant contributions were: the opening of membership to persons not eligible for membership in ACS; the sharing of technical information by his duties in technical sales as a supplier to the rubber industry; participated in sponsorship of local rubber groups, and in authorizing publication of *Rubber Chemistry and Technology*.

In the interview, Smith recalled, "The spirit of good fellowship is the hallmark of the division, evident nowhere else in the American Chemical Society." He further observed that the division's ability to work together for the common good of the rubber industry was one of the reasons the industry was so successful in building a massive new synthetic rubber industry under emergency wartime conditions.

The India Rubber Section's first decade was punctuated by growth, from both a numbers standpoint and a stature and recognition viewpoint. However, the United States was finally drawn into the "war to end all wars," and the patriotic India Rubber Section changed its course for a while. Decades later, in 1940, Ed V. Osberg, editor of *Rubber World*, authored a brief history of the Rubber Division published in *Rubber Chemistry and Technology*. He wrote, "Toward the end of the Section's decade of existence, the fruits of its labor were becoming evident. Analytical and test procedures were being perfected, and discussions on other subjects were becoming more frequent and livelier. The World War and rapid growth of the automotive industry had spurred technical activity in the rubber industry."

As Arnold Smith had reflected, indeed rubber scientists were making a difference.

PART FOUR

RUBBER WARS

IN THE
SHADOW

The great thing in the world is not so much where we stand, as in what direction we are moving.

OLIVER WENDELL HOLMES

In 1914, approaching the crescendo of the industrial age, the world paused to settle its differences. On June 28, 1914, when Archduke Franz Ferdinand of Austria was assassinated in Sarajevo, Serbia, and the killer was not granted extradition by Serbia to the Austro-Hungarian Empire, a series of delicate alliances were tested. In essence, during the seven-day period that followed, Germany declared war on Russia and France, and Great Britain declared war on Germany. For the time being, the United States maintained its isolation policy. In its period, the conflict was called the Great War, but many years later it would be renamed World War I.

World War I yielded many firsts. It became famous for trench warfare, the purposeful confinement of troops to holes in the ground for defensive reasons; the first large-scale bombing from the air; the first "industrial war" in history; and the first widespread use of chemicals as weapons of mass destruction. In addition, unlike previous wars, World War I displayed the importance of scientists and scientific knowledge to modern warfare. It also was the first opportunity for this unbreeched professional organization, the India Rubber Section of the American Chemical Society, and its membership, to display their resourcefulness and ingenuity. As William Woodruff, economics professor at the University of Illinois wrote, "The empiricism of nineteenth-century manufacture had also given way to the science of rubber chemistry. In no other country were these changes more pronounced than in the United States."[1]

Natural rubber, the resource that everyone had grown to love, was going to take a back seat for a while. It was during the India Rubber Section's first years that the concept of synthetic rubber was introduced when Fritz Hoffman received the first patent, coincidentally, in 1909, the section's birth year.[2] Hoffman worked for I. G. Farbenfabriken vorm. F. Bayer and Company, an organization that would later gain notoriety in the decades preceding World War II for the development of prototypical synthetic rubber products that exist today. Hoffman's research led to Bayer developing methyl rubber in 1911. He also demonstrated that, when combined with an appropriate amount of natural rubber, methyl could be used to make automobile tires.

There were several other efforts along the same lines being conducted throughout the United States, and from 1910 to 1912, David Spence and some colleagues at the Diamond Rubber Company in Akron, Ohio, performed some of their own tests on dimethylbutadiene.[3] Spence was born in Scotland in 1881 and educated at Royal Technical College in Glasgow and the Universities of Berlin, Jena, and Liverpool. He migrated to the United States after college and landed with Diamond as director of their research laboratories, but lower natural rubber prices forced their synthetic rubber research to be temporarily halted. In the post-prime of his life, Spence would become the first recipient of the Division of Rubber Chemistry's Charles Goodyear Medal in 1941.

The early years of the twentieth century, while filled with many sporadic and worthwhile research efforts in synthetic rubber, were not considered the premier era of its development. Spence's research and the efforts of others in the United States were halted between 1910 and 1912 because natural rubber became available at reduced prices. Hoffman's research with methyl rubber only led to some modest manufacturing efforts by Germany, and by the end of World War I, a total of only 2,350 tons had been produced. By comparison, in 1910, natural rubber from plantations had increased to 8,200 tons annually.[4]

In 1912, the India Rubber Section was barely walking. With its membership still below fifty men, it was operating in the very large shadow of the twelve-thousand-member American Chemical Society, but while an unprecedented level of amity had been reached with the ACS at that time, things would change with the outbreak of war. For one, World War I would challenge the patriotism—the allegiance to the countries of origin—of the members of the ACS and the Rubber Section. The conflict's beginning was enough to cancel plans for the American Chemical Society's fiftieth anniversary meeting in Montreal, Quebec, and split the membership—heavily spiced with German, French, and British chemists—if not physically, then certainly emotionally. Notices were even posted in surrounding clubs and other meeting places requesting that conversation about the war be avoided.[5]

In the meantime, back on red-white-and-blue soil, a prewar boom was materializing, and rubber companies, the major employers of Rubber Section members, were literally and figuratively treading on new roads. Car registrations had reached two hundred thousand by 1908, and this was enough of an impetus to turn on the dim, blackened factory light bulbs in Akron, Ohio.

Diamond Rubber wasn't one of the giants of the northeast Ohio rubber barons, but it had already given the industry David Spence, and two more of its top chemists also made a significant impact on the industry. In the mid- to late 1900s, Diamond's Arthur H. Marks developed an effective process for reclaiming rubber. However, one of his greatest contributions to rubber chemistry was probably when he hired his former

Harvard University classmate George Oenslager in 1905. Oenslager discovered the effects of organic accelerators on rubber processing, which enabled curing times to be slashed dramatically (see chapter 3).

At the Goodyear Tire & Rubber Company on East Market Street, founder Frank A. Seiberling had hired his first technically trained tire man, Paul W. Litchfield, to join the company. Litchfield, a graduate of Massachusetts Institute of Technology (MIT), designed tires, was a rubber compounder, ran the factory, and was head of personnel. In addition, as author Maurice O'Reilly noted in *The Goodyear Story*, Litchfield "even came to be known as the company doctor because his new office had a medicine cabinet on the wall with a first-aid kit—for restorative purposes—a bottle of whiskey."

Over at B. F. Goodrich (which would acquire Diamond in 1912) on South Main Street, there was never a more appropriate name for an individual than for Bertram G. Work, who had become president of the company in 1907. Work, too, was intent on keeping up with the demand for auto tires, but he also saw great opportunity in other rubber products and became an early leader in the manufacture of diversified rubber goods. He also wanted to ensure the company of a continuous supply of natural rubber, so he encouraged B. F. Goodrich to buy a plantation in Southeast Asia. Goodyear followed Goodrich's lead and also purchased a plantation, and on the other side of Akron, the Firestone Tire and Rubber Company was not far behind on its acquisition of a natural rubber plantation.

Firestone, one of the main Akron rubber giants, was led, ironically, by five feet six inches tall Harvey S. Firestone Sr. One of his competitors was feisty Irishman William F. "Will" O'Neil at General Tire. Both men were business savvy and manufacturing driven. That focus was understandable during an era when the plant was the center of the kingdom. Compared with the factory foreman, the role of the rubber chemist at the time was a rather subservient one. The chemist occupied most of his time with standardization of rubber analysis and physical testing. The manufacturing plants ruled with attitudes of "What can you do for me today? How can you help me make more black donuts?"

The tire manufacturers didn't confine their demands to their own researchers and chemists. Starting in the 1910s, the tire companies' penchant to cut costs precipitated developments of a few important technical innovations that helped to increase the efficiency of tire manufacturing. Three of those were the invention of the first commercially successful tire-building machine by Goodyear's William State, the flat band tire-building machine invented by U.S. Rubber's Ernest Hopkinson, and—likely the single most important advancement—the invention of the Banbury mixer by Fernley H. Banbury.[6]

Born in Cornwall, England, in 1881, Banbury left his native country in 1902 for America and Purdue University, where he received a B.S. in electrical engineering in 1906. His first job was with a German-based company, Werner and Pfleiderer, that manufactured kneading machines for the baking industry. Transferring the knowledge he received from twin-rotor batch kneaders, Banbury developed and patented a machine that would revolutionize the tire industry: the Banbury internal rubber mixer. Armed with his patent, Banbury joined the rubber industry via the Birmingham Iron Foundry in Derby, Connecticut, which merged with Farrel in 1916 to become Farrel-Birmingham. (Today it is the Farrel Corporation based in Ansonia, Connecticut.) Banbury knew he had a great product, but also knew he had to market it and sell it, skills he had not yet developed. In 1923, he looked toward Akron, Ohio, for help in peddling his machine.

Andrew Hale grew up in Akron in the cloud of the rubber industry. After receiving his degree in mechanical engineering from Cornell in 1916, he returned home to work in the engineering department of Miller Rubber Company. From 1916 to 1918, during World War I, he served in the U.S. Navy, and when the war ended he took a job with Firestone. In 1923, Fernley Banbury asked him for some help in marketing his product to the tire companies in Akron, and Hale became the sales office of Farrel-Birmingham with a mission of attracting the attention of the giants of the rubber industry.

"They didn't exactly beat a path to my door, and we only had about six machines in use at the time," Hale, one of the charter members of the Akron Rubber Group, recalled in a 1965 interview with Rubber Division historian Herbert A. Endres. "Our largest opposition came from technicians who thought the mixer wasn't strong enough to mix rubber and carbon black together." In spite of the large investment at the time (sixteen thousand dollars for a No. 11 machine without the motor), Hale finally convinced B. F. Goodrich, Miller Tire, and Goodyear of the merits of the mixer. Today, the Banbury mixer is a staple item in every tire plant in the world.

Fernley H. Banbury, who had joined the Los Angeles Rubber Group shortly after it was formed, became famous for his invention and in 1959 was awarded the Charles Goodyear Medal by the Rubber Division of the American Chemical Society. In later years, the division created a special award perpetuating Banbury's memory to honor innovations of production equipment widely used in the manufacture of rubber or rubber-like articles of importance.

Hale never received any known awards of national prominence, but he was proud of the fact that in 1810 his great-grandfather Jonathan Hale traveled twenty-eight days on horseback from Glastonbury, Connecticut, to Akron, Ohio, to establish a farm in the northern part of Summit County. In 1956, Hale's family bequeathed the entire 140 acres of land and property to the Western Reserve Historical Society and established Hale Farm and Village as a historical museum within the Cuyahoga Valley National Park. Today, Hale Farm and Village is a popular site for tourists with historic interests.

Ray Putnam Dinsmore, a young chemical engineering graduate from the Massachusetts Institute of Technology, also felt draped by the heavy cloak of an era focused on manufacturing. He earned his first job out of college at the Goodyear Tire & Rubber Company in 1914, when the technology of rubber was in its infancy, and Dinsmore, too, believed chemists still had several rungs to climb on the ladder of respect. "It was my opinion that the rubber industry depended far too much on empirical methods," Dinsmore said. "Its equipment was crude and its technical controls were completely inadequate. To remedy this, it appeared to me that a much broader exchange of information among rubber chemists was essential. There was, however, very little exchange except in connection with chemical and physical methods of analysis, and it was my constant effort to improve this situation. The chemist did not have much weight in the rubber industry, and he often tried to improve his status by being mysterious."[7]

While the American rubber companies were trying to meet the burgeoning demand from Ford's new motorized vehicles, the rest of the country was waiting to see, first, what happened in Europe with "their" war and, second, when or if the United States was going to join the hostilities. While America paused, the Germans were rounding up the usual suspects and cranking up their war machine, which was focused, in great part, on chemical arms production. In the middle of that was I. G. Farbenindustrie (or I. G. Farben), the cartel that included six dye companies: Badische Anilin and Soda Fabrik (BASF), Farbenfabriken vorm (Bayer), Farbwerke vorm (Hoechst), Aktiengesellschaft fur Anilinfabrikaten, Leopold Cassela, and Kalle and Company. In 1916, the companies merged into I. G. Farben, further enhancing Germany's war-waging capabilities and virtually controlling the new worldwide chemical industry. This included important controls over patents. This also sent up a warning flag that, for the time being, research on synthetic rubber would be placed on hold.

On April 6, 1917, slightly more than three years after World War I started in Europe, President Woodrow Wilson asked Congress for a declaration of war against imperialist Germany. Just six months prior to that Wilson had won reelection using the slogan "He kept us out of the war." Wilson really didn't care for the slogan, and he narrowly won both the electoral (277 to 254) and popular (9.1 million to 8.5 million) votes over Republican Charles E. Hughes, the U.S. Supreme Court associate justice.

In 1918, Lorin B. Sebrell had just completed his master's degree in chemistry at Ohio State University and decided he wanted to join President Wilson's new Chemical Warfare Service, directed by Maj. Gen. William L. Sibert of the U.S. Army. "We made some of the first toxic gases that were used in the war," Sebrell later recalled.[8] "At that time we had no particular protection. We would work until we couldn't see any longer in the laboratories, then go outside and sit underneath the trees until we cleared up enough to go back to work."

Sebrell was one of the first of fewer than eight hundred civilian chemists and twenty-two thousand military personnel to volunteer for the Chemical Warfare Service during World War I. He was also a member of the

India Rubber Section of the American Chemical Society. "In this work, one of the compounds was selenium analog of mustard gas," he said. "This did not have the vesicular action of mustard gas, but it did affect the optic nerve. I was the first chap who made this material, and I was in the hospital for a week without being able to see a thing. I even had to put my hands over my eyes on top of several layers of cotton bandage when the nurse opened the door [to check up on me]. It wasn't painful, but no one knew if I was ever going to be able to see again."

Sebrell indeed was able to see again, and his spirit and work ethic were typical of the rubber chemists' contributions to his country and the industry during those years. Born in Alliance, Ohio, on November 19, 1894, he had received his undergraduate degree, a B.S. in chemistry, from Mount Union College, and in 1922 returned to college to receive his doctorate in chemistry from Ohio State. He was hired by Goodyear's Dinsmore, and much later on in his career, in 1942, Sebrell would become the second recipient of the Division of Rubber Chemistry's Charles Goodyear Medal for his work with accelerators and antioxidants.

In September 1918 at a meeting in Cleveland of the Rubber Section, a committee on organic accelerators, which had been formed the previous year, reported on the toxic properties of the more commonly used organic accelerators. Division historian Benjamin Kastein wrote in 1984, "The Firestone study was quite extensive on the effects of 'hexa' (hexamethylene tetramine). Some persons were susceptible with effects noticed as skin rash and water blisters while others doing the same work were not affected. Repeated handling of uncured compounds in warm weather and contact with skin wet with perspiration caused the rash."

Reflective of the dichotomy of opinions and voting, and the United States' late entrance, World War I wasn't the catalyst for national unification that history declares World War II was. After all, the United States was only involved in the conflict for nineteen months. At 11:00 A.M. on November 11, 1918, the Armistice was signed. Nonetheless, research and development opportunities, while not plentiful, were available in several areas. The intrepid and pioneering efforts of the India Rubber Section and its chemists during this conflict, regardless of its length, provided a proving ground of sorts that bequeathed the seed for splinter groups, committees, and, eventually products.

The short period that the United States was officially involved in World War I is not a fair measure of the success (or lack of it) in the areas of rubber research and development. While the war, per se, didn't encourage an eruption of activity, it certainly didn't impede the progress. On April 7, 1919, at an American Chemical Society meeting in Buffalo, New York, the India Rubber Section became the Division of Rubber Chemistry, more or less controlling its own destiny, its own affairs, and its status as a nonprofit technical organization. Among some of the members who presented papers at that meeting was the young professor from the Municipal University of Akron, Hezzelton E. Simmons, who in 1913 replaced Charles M. Knight as director of the chemistry program. In 1928, Simmons became secretary and treasurer of the division, a position he would hold for seven years.

John B. Tuttle, chief chemist of The Firestone Tire and Rubber Company in Akron, Ohio, was the first chairman of the newly named Division of Rubber Chemistry and presided over the first meeting of the organization in Philadelphia on September 2, 1919. He also delivered a technical paper, "The Variability of Crude Rubber," describing organic accelerators to improve the low tensile strength of curing rubber. Tuttle, who had served the three previous years as secretary, served in his position only at the fall meeting, but several others who would be prominent in the division's early growth were elected officers. They included Warren K. Lewis of MIT and Arnold H. Smith of the National Bureau of Standards. In 1920 Lewis became the new division's first chairman to serve a full term. Smith handled the dual roles of secretary and treasurer.

Each played an important role in the development of the organization's structure. Because of his academic background and his role at MIT, Lewis provided the first element of a unique triple alliance between academia, industry, and government that would offer them a virtual road map for their future. With Smith's knowledge of the government, they were instrumental in developing objective professional relationships with the major rubber companies, encouraging them to submit technical papers to the meetings. In his first meeting as chairman, Goodyear's research group submitted six papers alone. Division historian Kastein wrote that Smith "probably more than any one person, was an important factor in positioning the Division for continued growth to its present status—such a contrast to the early dearth of papers and sparse attendance."

Smith was also influential in the American Chemical Society, establishing associate memberships as the division had initiated in 1920 under his leadership.[9] The original premise behind the formation of the ACS in 1876 was "the advancement of chemistry and the promotion of chemical research." In other words, membership wasn't restricted to professional chemists. Smith issued the suggestion to ACS secretary Charles L. Parson, professor of chemistry at New Hampshire College, and Parson agreed, provided that the associate dues would be twice the one-dollar fees for full members.[10]

The only challenges for Lewis and Smith and the rest of the division were to continue building on their achievements of "eliminating the secrecy" that permeated the industry in its first years and convincing the rubber companies that it was in their long-term interests to encourage their chemists and researchers to share information. In the end, they firmly believed that they would all benefit from such a controlled exchange of knowledge.

———————

In the history of the world, it is an unfortunate yet arguable premise that many monumental achievements and discoveries have been promulgated by wars or conflicts. Many products or services—too numerous to mention—received their designs because they were necessary to someone's or some country's pugilistic efforts. To a certain degree, the Division of Rubber Chemistry was born partly out of a national awakening to the value of chemistry as a science, and this movement was, by several accounts, a reaction to the Germans' dominance in research prior to and during World War I. At that time, scientific knowledge was funneled into a strategic part of modern warfare. In both present-day and past parlance, it was the patriotic thing to do.

Ironically, when the war ended, Americans welcomed the victory in much the same manner as a child would celebrate the last gulp of a tablespoon of castor oil. They wanted to distance themselves as quickly as possible from that conflict and any others that might occur in the future. They wanted to party, and the 1920s were ready to engage them. Later on in the century, the "Roaring Twenties" would be compared to the free-spirit decade of the 1960s. The similarities were jaw-dropping: The economy was prosperous, there was widespread social reform, new cultures were introduced, and people wanted to enjoy life and change their lifestyles. Like the 1960s, the 1920s were also good times for creative minds to expand.

Entrepreneurs were running rampant in the 1920s rubber industry, but so were the politicians. Between 1920 and 1928, rubber prices would fluctuate seemingly like the winds in a hurricane from the eye to the rain shields, rising to $1.13 a pound from less than 3 cents.[11] To combat the volatility, the British government introduced the Stevenson Plan (which was repealed in 1928) to limit output in the new plantations of its colonies in Malaysia and Ceylon to stabilize prices. This had an obvious adverse affect in the United States, prompting U.S. Secretary of Commerce Herbert Hoover in 1925 to declare the plan "a threat to world peace."[12]

Just two years earlier in 1923, Harvey Firestone had convinced Congress that rubber companies should have their supplies under their control. Congress agreed and five hundred thousand dollars was appropriated for studies of all possible rubber-producing areas in the world to determine the feasibility of establishing U.S.-owned plantations. Then Firestone quickly inaugurated his rubber plantations in Liberia. Automobile magnate Henry Ford followed his friend Firestone's lead and partnered with inventor and botanist Thomas A. Edison to establish a plantation in Brazil.[13]

The fluctuating markets also took a toll on Goodyear and Frank A. Seiberling, who founded the enterprise and ran it for its first twenty-three years. He resigned from the company in 1921 and formed Seiberling Rubber. In addition, other fellow executives resigned, including Arnold H. Smith. Smith, who had served with the Bureau of Standards until he joined Goodyear in 1919, formed the Rubber Service Lab in Akron, which was later absorbed by Monsanto in 1929.

Regardless of the political maneuvering and the market fluctuations, the rubber industry and members of the Division of Rubber Chemistry continued to grow and progress. Ray P. Dinsmore, hired by Paul W.

Litchfield at Goodyear in 1914, noted some achievements of the early 1920s in an article in *Industrial & Engineering Chemistry* in 1951: "From 1920–24, (Clayton W.) Bedford (Loren B.) Sebrell and collaborators studied ultra and semi-ultra accelerators . . . much more powerful than their predecessors which made a practical pneumatic truck tire possible and added materially to the life and quality of passenger tires and other products."[14]

There were many achievements that were significant to the technical growth of rubber during the 1920s, and just a few mentioned by Dinsmore were the Geer oven-aging test and the (John M.) Bierer— (C. C.) Davis oxygen bomb;[15] compounds to inhibit oxidation of rubber by Herbert A. Winkelmann and Harold Gray; Sidney M. Caldwell's introduction of condensation product of acetaldehyde and aniline; and G. Stafford Whitby's extensive research in several areas of rubber chemistry. However, business and entrepreneurship were dominant for most of the decade.

The multiple developments and subsequent successes of niches of technological research further emphasized the need for sharing of information among the members of the Division of Rubber Chemistry. In 1926, although the American Chemical Society had allowed the division to meet separately once a year, the increasing desire to exchange data more often would soon yield the birth of a key element in the organization's overall success and effectiveness.

The highest honor bestowed by the American Chemical Society's Rubber Division, the Charles Goodyear Medal, is named for Charles Goodyear, who discovered the vulcanization of rubber in 1839. (Rendering of Goodyear from the *Scientific American Supplement*, No. 787, January 31, 1891.)

Charles Cross (C. C.) Goodrich, pictured in this photo from St. Paul's School, Concord, New Hampshire, was the first chairman of the India Rubber Section of the ACS in 1909. Goodrich was the oldest son of Benjamin F. Goodrich, founder of the B. F. Goodrich Company, and attended St. Paul's School from 1885 to 1889. (Photo courtesy of the University of Akron Archives.)

The first secretary-treasurer of the Rubber Section in 1909 was Frederick J. Maywald, owner of several patents. In 1910, Maywald also was a member of New York City's Municipal Explosives Commission, which investigated the December 19 explosion at the city's Grand Central Power House. Ten people were killed instantly and 150 people were injured. The *New York Times* wrote, "Probably no serious catastrophe in the city's history has been so thoroughly investigated." (Photo published in *India Rubber World*, February 1, 1910.)

In 1919, Firestone's John B. Tuttle was elected the first chairman of the new Rubber Division. A graduate of the University of Pennsylvania, Tuttle left Akron and Firestone two years later and headed to New York to go into business for himself as a consultant on rubber chemistry. He later held positions on the U.S. Navy Department's Committee on Rubber Goods Specifications and the Joint Rubber Insulation Committee. (Photo published in *India Rubber World*, January 1, 1921.)

Rev. Julius A. Nieuwland, C.S.C., teaches at his chemistry laboratory at the University of Notre Dame, South Bend, Indiana. Nieuwland's work in acetylene chemistry pointed the way for DuPont's development of Duprene, the first commercially successful synthetic rubber. On November 2, 1931, Nieuwland and a team of chemists from DuPont announced their findings at a meeting of the Akron Rubber Group at the University Club. Nieuwland also was mentor to a young graduate assistant named Knute Rockne, who later became a head football coach and a Notre Dame legend. (Photo courtesy of *Rubber & Plastics News.*)

William J. Sparks, a chemist at Exxon, was one of only three men who served as both chairman of the Rubber Division (1960) and president of the American Chemical Society (1966). In 1937, Sparks collaborated with Robert Thomas in developing a workable butyl rubber. He loved golf and held several patents for improved grips for golf clubs. (Photo courtesy of the University of Akron College of Polymer Science.)

Another heralded Exxon chemist, Robert M. Thomas, 1969 Charles Goodyear Medal recipient, worked with William J. Sparks to develop butyl rubber. In their honor, the Rubber Division established the Sparks-Thomas Award in 1986 supported by ExxonMobil Chemical Company. (Photo courtesy of the University of Akron College of Polymer Science.)

In the manufacture of anything that includes rubber, the Banbury mixer is a staple item. In 1916, Fernley H. Banbury invented the machine that bears his name. In 1959, Banbury was awarded the division's Charles Goodyear Medal. Today, the Farrel Corporation sponsors the division's Fernley H. Banbury Award honoring innovations of production equipment widely used in the manufacture of rubber or rubber-like articles of importance. (Photo courtesy of the Farrel Corporation).

Studies by Melvin Mooney on the flow of unvulcanized rubber resulted in his invention of the Mooney shear viscometer, a testing instrument that is used worldwide in the rubber industry. Mooney was recipient of the Charles Goodyear Medal in 1962, and the division initiated the Melvin Mooney Distinguished Technology Award (sponsored by Lion Copolymer) in his honor in 1982. (Photo courtesy of AIP Emilio Segre Visual Archive.)

Hezzelton E. "Hez" Simmons, secretary-treasurer of the Rubber Division from 1928 to 1933 and Charles Goodyear Medalist in 1952, signs diplomas as part of his full-time job as president of the University of Akron. He served in that capacity from 1933 to 1951. (Photo courtesy of the University of Akron Archives.)

David Spence, a research chemist from Scotland hired by Diamond Rubber Company (later absorbed by B. F. Goodrich), was the first recipient of the Charles Goodyear Medal in 1941. Spence (*far left*) received the medal primarily for his research in guayule. Next to Spence are Maj. Evan W. Kelley of the U.S. Forest Service, in charge of the government's guayule project, and Senator Sheridan Downey (D-Calif.). (Photo courtesy of the Bancroft Library, University of California, Berkeley.)

Harry Fisher holds a cross-section of a tire at his desk at the University of Southern California in 1954. Fisher was Rubber Division chairman in 1928 and president of the American Chemical Society in 1954. (Photo Courtesy of the University of Southern California Archives.)

Waldo Semon was a key member of the B. F. Goodrich team of chemists who were actively involved in the Synthetic Rubber Research Program during World War II. Appropriately, Semon (*second from right, without a coat*) was the division's Charles Goodyear Medalist in 1944. In 1952, he was elected division chairman. (Photo courtesy of the University of Akron Archives.)

The first gathering of the 25-Year Club was held at the Sherman Hotel in Chicago in 1948.

TECHNICALLY
SPEAKING

*The saddest aspect of life right now is that
science gathers knowledge faster than society
gathers wisdom.*

ISAAC ASIMOV

In 1928, penicillin was invented, vitamin C discovered, the
animated (black-and-white) movie *Steamboat Willie* introduced a cartoon character named Mickey Mouse, bubble
gum (a derivative of rubber)[1] was perfected, and, receiving
a considerably lesser amount of international acclaim, the
journal *Rubber Chemistry and Technology* was founded.
For any scientist, chemist, or researcher who realizes the
importance of his or her discovery, the value of commu-
nicating those results to their colleagues, their peers, is
almost as vital as the discovery itself. Publication is vital to

the process of sharing information with colleagues, and nothing does it better than a scholarly journal.

In June 1967, John McGavack, who had been instrumental in the maintenance of the now-defunct *Bibliography of Rubber Literature* from 1940 to 1964, conducted an exhaustive study of the top one hundred contributors to the world's rubber literature between 1932 and 1966. His results were published in *Rubber Journal* and *Rubber Chemistry and Technology* and included technical articles published in the twenty-year time frame from authors in twenty-six nations. Likely to no one's surprise in the rubber industry, *RC&T* was the preferred forum for the more than twenty-two thousand references.

A similar study was conducted in December 2000 by Wai Sin Tiew and K. Kaur and published by the *Malaysian Journal of Library & Information Science*.[2] The Malaysian study yielded similar results, and *RC&T* again topped the list as the most cited journal or serial publication in the field of rubber research. The study further stated, "In terms of frequency distribution of journals/serial citations, most of the serial publications (95.07%) received between one to ten citations. *Rubber Chemistry and Technology* was the most cited journal/serial title with 198 citations." It concluded, "Hence, this conforms to earlier studies, which indicate that rubber scientists depend highly on journals/serial literature."

Arnold H. Smith would have been proud. The Rubber Division chairman of 1929 knew back in 1920 the importance of technical publishing. At the division's Chicago meeting that year, he posed the question, "Shall the Rubber Division publish an annual volume of reprints and abstracts of everything of interest to the rubber chemist made public during the year?" The seed for *Rubber Chemistry and Technology* was planted.

In the same year and in the same meetings that created the local rubber groups, *RC&T* was established. At a meeting of the executive committee in Akron on November 14, 1927, Carroll Campbell "C. C." Davis (of Bierer-Davis oxygen bomb fame) of Boston Woven Hose and Rubber Company, was elected the first editor with the authority to select his own associate editors. Walter W. Evans of Goodyear, division chairman in 1921, was elected managing editor. Annual division dues were increased to two dollars for members and four dollars for associate members to underwrite the expense. In April 1928, shortly after the new local rubber groups were formed, Davis published volume 1, number 1 of *RC&T*, containing twenty reprinted papers and abstracts. Six papers related to rubber derivatives and hard rubber, five were on reclaimed rubber, five covered aging and vulcanization, and four involved testing analysis of rubber.

In its December 1966 special edition on the division's history, *Rubber World* editors mentioned that Davis was fluent in several foreign languages and added, "Davis created an industry-wide group of advisors and readers and engaged in extensive correspondence regarding the articles and subjects he selected. As the years passed, the journal came to occupy more and more of his spare time—and even pulled his wife into its vortex. The two of them turned their home into an editorial sanctum, packed with references, papers, files and the accumulating volumes that appeared under his direction."[3]

In a 1966 interview by division historian Herbert A. Endres of John M. Bierer, Davis's former boss at Boston Woven Hose, Bierer said of Davis's contributions, "He was a fine man and a hard worker. Davis always had the full support of our company because for what he gave, we received much more in return. We were kept up to date on the technical issues, better than most companies. I'm not a religious man, but I certainly believe what the Bible says is true: Cast your bread on the surface of the waters, for you will find it after many days."[4] In 1955, Davis was awarded the Charles Goodyear Medal. He was editor of *RC&T* for twenty-nine years until his death in 1957.

Each of Davis's successors was also extremely technically competent. David Craig of B. F. Goodrich was Davis's associate editor for two years and took over in 1958. Craig began lab work under Waldo L. Semon and later worked with the World War II Synthetic Rubber Program. While working at his laboratory in Brecksville, Ohio, on July 16, 1964, Craig suffered a heart attack and died at age fifty-nine.

In 1964 Edward M. Bevilacqua, Philadelphia native and Ph.D. graduate of the University of Wisconsin, was editor and served four years representing Uniroyal. Early C. Gregg Jr. became the fourth *RC&T* editor in 1969, the second editor from B. F. Goodrich. The Boston native and Harvard graduate was the first editor to also serve as division chairman

(1976). In 1975, H. Karl Frensdorff from DuPont was the first foreign-born editor. A native of Hanover, Germany, he received his Ph.D. in physical chemistry from Princeton in 1952 after serving in the U.S. Army during World War II and becoming a U.S. citizen in 1944. Aubert Y. Coran of Monsanto became the fifth editor of *Rubber Chemistry and Technology* on January 1, 1978, and in 1995, Coran became the second editor to receive the division's prestigious Charles Goodyear Medal.

The importance and value of *RC&T*'s first papers was manifested by fourteen of those articles being reprinted in *Industrial & Engineering Chemistry*, two each from *Rubber Age* and *Berichte der Deutschen Chemischen Gesellschaft* and one each from *India Rubber World* and *Institution of Rubber Industry Transactions* in London. Recapping the first fifty years' history of the journal in 1978, Goodyear's Tom H. Rogers wrote, "The subject matter of the papers represents a good indication of the scientific interests of the industry at that time; namely, how to put natural rubber to new and various uses, reclaiming used and abandoned rubber products, improvement of vulcanization and aging, and the evaluation and analysis of rubber and rubber compounds."[5]

The first edition of *RC&T* contained 182 pages and for the first year had a total of 594 pages in four volumes. The journal retained its quarterly publication format until 1957, when it began issuing five volumes a year, the standard maintained today (there were a few years when *RC&T* published six times a year). Also in 1957, a committee appointed by division chairman Arthur E. Juve the year before recommended the establishment of *Rubber Reviews*, an annual publication. Traditionally in the third issue of every year, *Rubber Reviews* offers comprehensive overviews of rubber-related topics that are normally longer (on average thirty to fifty pages) than the typical technical papers.

GROUP THERAPY

6

The Division of Rubber Chemistry is unique in the very fact that it sponsors the Rubber Groups. These provide a common ground between the chemists and other men in the industry—salesmen, engineers, suppliers— even interested bystanders.

WILLIAM J. SPARKS, CHAIRMAN
OF THE RUBBER DIVISION, 1960

At the start of the 1920s, the Dow Jones industrial averages hovered around the 60 mark. By the first of October in 1929, the market index had shot up to a record 400.[1] But the euphoria of the 1920s was given a death sentence on October 28, 1929, when the averages plummeted and overnight millionaires suddenly became bankrupt. Even many small investors, who had entrusted their money to banks, lost their life savings as the United States was thrown into the Great Depression.

While not a driving factor, the Depression likely hastened and influenced the Division of Rubber Chemistry to review travel costs (car, bus, train, airplane, hotels, meals, etc.) to members when selecting meeting sites. It was also a subliminal message that if they wanted to increase attendance, especially during tough times, maybe they should consider "satellite" meetings. After all, just a few years before, in 1926, the American Chemical Society had cancelled its proposed meeting in Los Angeles because prospects for good attendance were poor.[2]

In 1921, Ray P. Dinsmore had just returned from Los Angeles and a temporary assignment with Goodyear and was anxious to reacclimatize himself with fellow chemists, realizing that the division offered him "the opportunity for the practical exchange of information with regard to rubber chemistry and physics which I had been looking for . . . I lost no time in becoming associated with the division."[3] Dinsmore knew the value of exchanging information with professional colleagues. He even became instrumental in helping to form the Akron Section of the American Chemical Society in 1923 so that Akron members "wouldn't have to drive all the way to Cleveland" to attend a meeting.[4]

In 1927, Dinsmore became chairman of the Rubber Division, and on April 13 at the spring meeting in Richmond, Virginia, that year, the executive committee suggested that they consider formation of related rubber groups—geographically distributed—similar to the sections of the American Chemical Society. There was also discussion about creating a technical journal to communicate the increased number of papers being presented at the meetings (see chapter 5). At the fall meeting in Detroit on September 8, 1927, a formal proposal was issued that groups be formed in four cities and that membership be open to technologists in the rubber industry who were not chemists. Harry L. Fisher, who became division chairman in late 1927 (calendar year 1928), appointed a committee, by respective geographical region, to organize such an endeavor. The committee members were past-Chairman Dinsmore, Akron; C. R. Boggs (1925 division chairman), Boston; A. A.

Somerville, New York; and R. B. Stringfield, Los Angeles. The proposal was approved, and the first four local rubber groups were almost an immediate success.[5]

On January 11, 1928, the New York Group was the first to meet, and 240 attended and elected W. A. Gibbons as chairman and D. F. Cranor as secretary-treasurer. The Akron Group met on February 15, 1928, with more than 300 "chemists and engineers attending," and elected Herbert A. Winkelmann of General Atlas Carbon as chairman, W. H. Fleming as vice-chairman, and R. J. Bonstein as secretary-treasurer.[6] The Los Angeles Group met for the first time on May 11, 1928, and 43 attendees elected organizer Stringfield as president, A. K. Pond as vice president, E. S. Long as secretary-treasurer, and C. R. Park, Long, and Pond to the executive committee.[7] On November 7, 1928, the Boston Group was the fourth to meet, and J. M. Bierer was elected chairman and T. M. Knowland secretary-treasurer.[8]

The following year, the Chicago Group was the first to join the initial foursome and, on June 28, 1929, formed under the leadership of Chairman C. Frick, Vice Chairman Otto Urech, and Secretary-Treasurer B. W. Lewis. The first official meeting of the group was held on October 25, 1929, in conjunction with a meeting of the Chicago Section of the American Chemical Society. It was eight more years until Detroit became the sixth local rubber group on June 23, 1937. Seventy-five individuals selected W. J. McCourtney as chairman, E. J. Kvet as vice chairman, H. C. Anderson as chairman of the membership committee, and J. H. Norton as entertainment committee chairman. Ed V. Osberg, editor of *India Rubber World*, wrote in his "History of the Rubber Division" in the 1940 edition of *Rubber Chemistry and Technology*, "The work of these local groups is well known to all in the rubber industry. They afford an opportunity for supplementing and extending the technical activities of the Division and for those present from the rubber and allied industries to become better acquainted so as to encourage a spirit of cooperation and thus increase the benefits possible from the exchange of ideas."

Since 1940, rubber groups have experienced highs and lows and have consolidated or merged, but regardless of their present status, history reveals that most rubber groups have had some common threads, and outside activities intended to build camaraderie and community involvement have been prominent. For example:

1. The West Michigan Rubber Group counts fishing and golf outings as two of their social events while they actively support one of the few student chapters of the Rubber Division at nearby Ferris State University.
2. Organized on February 14, 1946, the Connecticut Rubber Group (CRG) was formed to "specifically focus on providing technical information to those who would not normally be able to attend meetings at remote locations." In 1965 and 1966, they held a course in "Engineering with Elastomers" at Southern Connecticut University and in 1967 joined with the Rhode Island and Boston Rubber Groups for a New England technical meeting. Those meetings are still held every two years.
3. The Twin Cities Rubber Group in the Minneapolis-St. Paul area was formed on March 14, 1963. Its first president was Alex Kaminski of Houghton Vix-Syn Company.

At their peak in 1984, there were twenty-six active local rubber groups with a total membership (including two Canadian groups) estimated at almost eleven thousand.[9] With deactivations, consolidations, and mergers, there were eighteen active groups as of 2009. In 1965, the Akron Rubber Group grew so large that it became the first local to qualify for a second area director on the Division of Rubber Chemistry's Board of Directors by having more than 750 members. Peak membership for the Akron group was in 1970 with 2,226 members.

One of the most recent additions to the portfolio is the Energy Rubber Group founded in 1981 by Fred Pedersen, a former B. F. Goodrich technical service manager and a 1973 graduate of the University of Akron. "I went to the best-known polymer school in the country and got a degree in marketing," he said, laughing. The Energy Rubber Group was formed like most groups, out of desire to share information. It was also the first nongeographic group. "We were founded out of a need to focus on applications for organizations that service the energy industry," said Pedersen, founder and president of Production Systems, Inc., a consulting firm to those industries, and the group's first chairman. At the first informational meeting of the group in 1980 at a hotel in Irving, Texas, twenty-eight people showed up. "I asked them for five dollars each to cover the cost of the room," he said. "We just broke even." However, at their first official organizational meeting in 1981, more than 180 attended, including Lucian Samples, and "when I saw him walking through that door," said Pedersen, "he gave us instant credibility."[10] Samples, a vice president of Murray Rubber in Houston, had been a long-time member of the Southern Rubber Group and a well-respected rubber industry veteran in the southwest.

Today, the local rubber groups are fairly autonomous and, reflective of the mature rubber industry, experience many of the same growth issues. However, Rubber Division executive director Ed Miller believes that

> The history of the Rubber Division ACS closely parallels the history of the rubber industry and in part the world. For a century, the division has been continuously evolving and growing in its services and programs that serve our members and their organizations. This was possible only because we stood in the light of visionaries and giants. We have a long tradition of self-sacrifice and dedicated contributions by volunteers who first had the vision and then helped to build this association.
>
> Their monumental efforts and persistence to advance the division and the rubber industry were catalysts. These contributions of time and work are amplified by the exceptional support of the professional staff that worked daily to help make the division's activities successful and provide value to the rubber industry and the professionals within it. The Rubber Division continues to provide the technical, educational, business, and networking resources for the industry, working hand in hand with the University of Akron, the American Chemical Society, other professional associations, and the many companies that are integral to this country.
>
> Following this historical first century, the division will continue to serve society as we follow our vision to "Enhance Science, Technology, and Business Across the Evolving Elastomeric Community."[11]

George K. Hinshaw considered himself just a farm boy from Pontiac, Illinois, in the heart of corn country. He had received his B.S. in chemistry in 1913 from Illinois Wesleyan in nearby Bloomington—just a spell down the road—and enjoyed his time in college with his fraternity brother Carl Marvel, who he "juggled test tubes with" in the labs on occasion. Hinshaw thought he had his dream job when he was offered a position teaching chemistry and coaching football at Township High School in Pontiac. He could also run the eight-thousand-acre family farm and finish his master's degree at his alma mater, only thirty-five miles away. In 1915, he received his master of science degree, but he was running short on money to repay his college loans. "I hadn't seen my sister in over a year and a half and I really wanted to see her," he recalled. "She and her husband had moved to Akron, Ohio, but I didn't have enough money to visit her. She told me about the rubber companies there and suggested that I write to them about a job. So I did. I wrote to all of them—literally pleading for a job. Well, I got back four offers, and then I had to decide which one. I liked the sound of Goodyear. When I went to visit my sister I interviewed at Goodyear, and in 1917 I accepted the job."[12]

Hinshaw began as an analytical chemist, but in six months "they yanked me down to Plant One as an assistant to Dinny [Ray P. Dinsmore] and I worked for him until he went to the Synthetic Rubber Program during World War II." Since Hinshaw was accustomed to working many hours a day with a farm, coaching, and teaching, he also joined the Division of Rubber Chemistry and a few years later became a charter member of the Akron Rubber Group, serving as its second chairman in 1929. In 1931, Hezzelton E. Simmons, professor of chemistry at the University of Akron, who was also the division's secretary-treasurer, went to Hinshaw with an unusual request. "Hez asked me to call a meeting of the Akron Rubber Group in secret for something 'I would never regret,'" Hinshaw said. "I didn't know what it was, but I had to use all my showmanship and some deceit to get a large attendance at this meeting. So we called the meeting—I guess on some hallucinations—and we had a full house."

The meeting that Hinshaw was organizing for the Akron Rubber Group would develop into a most historic event and, under the auspices of the division, would serve as a launching platform for events at future meetings. Up to this point, the local rubber groups were like spare tires to the division's automobile. However, one of those tires would rotate to the drive axle in 1931.

DIVINE
DISCOVERY

*The best thing I ever learned in life was that
things have to be worked for. A lot of people
seem to think there is some sort of magic in
making a winning football team. There isn't,
but there's plenty of work.*

KNUTE ROCKNE,
CHEMISTRY PROFESSOR AND FOOTBALL COACH,
UNIVERSITY OF NOTRE DAME

Arnold M. Collins was an adventurer who liked to experiment with things. One spring day in April 1930, he was working with a chlorine compound in his laboratory at DuPont in Wilmington, Delaware. He already had experienced his share of lab explosions, but what came next may have been a *Eureka!* moment in chemistry even tantamount to the day in 1839 that Charles Goodyear spilled one of his sulphur-laced concoctions onto a stove and accidentally discovered vulcanized rubber. "I put this stuff in one of these shaking bottles," Collins recalled years later, "and started shaking it. Julian Hills [an associate] in the laboratory said, 'What are you doing?' Well, I said, 'I don't know, but it's heating up!' So we put our hands on this bottle that was in the shaking machine and I said, 'Let's get out of here!'"[1]

Neither Collins nor his associate was injured—then. The lab didn't blow up because they had the wisdom to put the beaker in water to cool it off before anything drastic occurred. However, what happened in that DuPont lab—accident or not—actually was the culmination of years of research by several people that led to the eventual discovery of the first high-quality, usable, and marketable synthetic rubber: Duprene, today called Neoprene. The new product, with the generic name of poly-chloroprene, was first announced to the world on November 2, 1931, at a meeting of the three-year-old Akron Rubber Group (that meeting would also be the sounding board for a future Charles Goodyear Medalist).

South Bend, Indiana, is about 278 miles from Akron, Ohio, or 664 miles from Wilmington, Delaware, via today's interstate highway system. In the early part of the twentieth century, South Bend was a typical mid-American city. It was a factory town—similar to Akron or Wilmington in that era—with prominent U.S. companies that proudly carried names like Studebaker, Ford, and South Bend Street Railway Company, just to name a few. Even during World War II, the South Bend Blue Sox All-American Girls Professional Baseball League team was formed. It was active from 1943 to 1954.

When someone mentions South Bend today, most people think of one thing—the University of Notre Dame. And whether it's folklore, fantasy, fiction, or fact, the college has built a solid reputation around football. Just mention Knute Rockne, Frank Leahy, the "Gipper," or Joe Montana and either Pat O'Brien, Ronald Reagan, or Notre Dame come to the frontal lobes. Football notwithstanding, Notre Dame was placed on the northern Indiana map in 1844 by a twenty-eight-year-old French priest, Reverend Edward Sorin, and seven of his colleagues, all of them members of the recently established Congregation of Holy Cross. The Congregation of Holy Cross was founded in Le Mans, France, in 1837 by Venerable Basil Moreau to administer to the educational and pastoral needs of the Catholic Church in France. Father Sorin obviously took the order's mission as a personal goal and founded Notre Dame to pursue those ideals.

Forty-one years later, an individual was born who would later become a revered member of that order, a distinguished member of the faculty of the University of Notre Dame, and literally the catalyst to one of the twentieth century's most compelling discoveries, a viable alternative to natural rubber. His colleagues in this venture would later use Akron, Ohio, as a launching board for this new discovery, and it would eventually earn him a spot in the National Inventor's Hall of Fame. His name was Julius A. Nieuwland, a.k.a. Father Nieuwland.

Nieuwland was born on February 14, 1878, of Flemish parents in Hansbeke, Belgium, and immigrated when he was two years old with his family to the United States and South Bend, Indiana. Nieuwland grew up in South Bend and eventually graduated from Notre Dame in 1899, enrolled in the seminary shortly thereafter, and was ordained a priest in 1903. He received his Ph.D. from the Catholic University of America in Washington, D.C., in 1904, and then took his doctorate back to his undergraduate alma mater to teach and in 1918 he became a professor of organic chemistry.

It was during this time that Nieuwland's research focused on acetylene chemistry (he was much later called the "Father of Acetylene Chemistry"), and in his work he produced divinyl acetylene (DVA), a jellylike material that transformed into an elastic compound similar to rubber when passed over sulfur dichloride. In 1925, Nieuwland's interest in acetylene prompted him to write a paper on his study of the hydrocarbon, "Acetylene Reactions, Mostly Catalytic." In this paper, Nieuwland described his previous work, which yielded the raw rubbery material that then polymerized rapidly at room temperature to a hard resin. Unfortunately, this also caused an unwanted result: The material would sometimes explode when struck.

About the same time that Nieuwland was working on refining his discovery, Elmer K. Bolton, DuPont's director of research on organic chemistry, who in 1913 had completed postgraduate work on synthetic rubber at the Kaiser Wilhelm Institute in Germany, also had a keen interest in other alternatives to natural rubber. After all, the Germans had been working on the development of a synthetic rubber as early as World War I and had developed a well-established organic chemical and substantial rubber chemicals industry. During this time, most fundamental research was conducted in universities, but polymers and polymerization were not yet well understood.

Bolton had worked on a variety of mixes and alternatives to the Germans' compound and wanted to learn—not from their successes but from some of their less-than-successful attempts. He grasped at every opportunity to learn more about synthetic rubber, and on December 29, 1925, the Division of Organic Chemistry sponsored its first National Organic Symposium in Rochester, New York. The symposium was chaired by Rev. Julius Nieuwland, C.S.C. of the University of Notre Dame, and the secretary was Frank C. Whitmore of Northwestern University in Evanston, Illinois.[2]

Nieuwland delivered his paper on a highly unsaturated derivative of acetylene and its possible ultimate value in the development of synthetic polymers. Bolton was in the audience and was intrigued with Nieuwland's report, believing it had great potential in his work and that divinyl acetylene might be used as the basis of a new rubber and the reaction might also offer the more desirable alternative of making monovinyl acetylene, a potential raw material for butadiene.[3] Bolton asked for and was granted a meeting with Nieuwland and learned that the priest had abandoned his work with divinyl acetylene because of its volatility. Nieuwland, however, agreed to allow DuPont researchers access to his findings to see if they could develop a more stable polymer.

Bolton went back to DuPont and asked his company to draft an agreement with Father Nieuwland to serve as a consultant during their work. Nieuwland liked the idea of working with industry, but said he couldn't collect a fee because of his Holy Cross vow of poverty. DuPont agreed to earmark Nieuwland's monthly retainer of one thousand dollars plus royalties to the University of Notre Dame for library books.[4] In the meantime, Bolton enlisted the help of Wallace H. Carothers, who had come from Harvard in 1928, and Arnold M. Collins to investigate all potential uses for Nieuwland's capricious development.

Work on Nieuwland's premises began in earnest in January 1930. Bolton asked Carothers to examine the chemistry of an acetylene polymer with the objective of creating a synthetic rubber. In April, Collins, who had been experimenting with synthetic resins in another department, successfully isolated chloroprene, a liquid that polymerized to produce a solid material resembling rubber. He had the advantage of some prior work of another DuPont team, which included W. S. Calcott, F. B. Downing, and A. S. Carter. The trio had completed a group of experiments with divinyl acetylene, but none gave them synthetic rubber. However, they discovered an important clue that led to Duprene's ultimate success. It was a more active monovinyl form of the compound (with one double-bond) that consistently accompanied the divinyl form and could be produced in sufficient yields.

Collins's success came when he was away from the lab. On a Friday afternoon in April 1930, his team devised an old-fashioned still to conduct their purification experiment. Over the weekend, something strange happened. By Monday, the low-boiling liquid fraction, which had been collected in an attached test tube, had congealed—not to a resin, but to something that they had never seen before. The experiment yielded a product with a lively bounce and other characteristic physical properties of natural rubber. By accident, Collins had discovered a new type of synthetic rubber. With a few more refinements, including getting the new polymer into a form that could be tested, the team had developed a prototype of chloroprene, as Carothers named it.[5]

In short, Nieuwland's successful polymerization of acetylene into divinyl acetylene enabled the DuPont team to focus on monovinyl to react the substance with hydrogen chloride gas to develop chloroprene and Duprene (Neoprene), the first widely used synthetic rubber. "Practical methods for making plastic, millable polymers of chloroprene [the neoprenes] were next developed," said Collins. "The first being made by the method of Ira Williams [Rubber Division chairman in 1934 and the fourth recipient of the Charles Goodyear Medal in 1946] involving the partial polymerization of the chloroprene to a plastic, soluble polymer and removing the unchanged monomer, as in the process for making the soluble synthetic drying oil from divinyl acetylene."

In the fall of 1931, natural rubber (#1 ribbed smoke sheet) was selling for less than five cents a pound. Who in his or her right mind would listen to a discussion about an alternative to natural rubber that was estimated to sell for about one dollar a pound? The answer: individuals and chemists with extremely open and creative minds.

In the late 1920s, after he had visited Nieuwland, and his team at DuPont had progressed with their research in chloroprene, Bolton decided to seek the advice of one of the nation's top experts in the area of rubber chemical research. "Dr. Bolton went out to Akron, Ohio, to talk with Dr. [William C.] Geer," said Ernest Bridgewater, who in 1931 was director of the rubber chemistry division and director of sales for DuPont. "Web Jones was a bosom pal of Bolton's who introduced him to Dr. Geer and Bolton also knew Geer through a research organization they belonged to that met in New York every few months. He came back from Akron with not too much encouragement, but he had the support of the president and chairman, Mr. Lammot Du Pont, who was very keen about getting the company into other businesses besides lines of explosives."[6]

Bridgewater surmised that Geer told Bolton that the proposition, while worthy, might not be financially practical at that time with the anticipated price of one dollar a pound for the price of natural rubber at the time. One of Geer's colleagues at B. F. Goodrich, Harlan L. Trumbull, however, had some inside information. As Trumbull told division historian Herbert A. Endres in 1965,

In 1926, at a Rubber Division meeting in Philadelphia, Geer had given a pessimistic report on synthetic rubber based on petroleum. He cited the possibility of a more lucrative research being to distill natural rubber as a supply of lubricants superior to those derived from petroleum. This startling suggestion was consistent with the negation of the proposal made in 1919 that the company undertake research on synthetic rubber. Parenthetically, Bolton in about 1929 asked to be permitted to visit B. F. Goodrich in Akron to familiarize himself with the essential properties of rubber in industry. He confided in a few of us that his company [DuPont] had left it up to him whether they should spend a half-million dollars on research leading to synthetic rubber. This, of course, referred to the great Duprene project that gave birth to Neoprene.[7]

But DuPont "thought it had a winner, and decided to go through with it," said Bridgewater, who would become Rubber Division chairman in 1932. The reaction of the DuPont sales team was predictable. "We were enthusiastic as hell," Bridgewater said. "By that time we knew the product had good ozone and oil resistance . . . and the fact that it was ozone resistant was particularly appealing to the tire companies that had experienced a great deal of tread cracking and sidewall cracking . . . we knew we had a saleable product."

Other applications for the new product that Bridgewater envisioned were for a variety of "mechanical goods" products, such as conductors in between the spark plugs and distributors in a car's engine, gaskets, oil seals, and suction and discharge hoses that demanded superior oil-resistance capabilities. "The standard types of industrial hose at that time were lined with bare fabric held in place with wire," he said. "Neoprene would enable the manufacture of the hose with the wire inside of the rubber."

November 2, 1931, fell on a Monday, a day before the nation would go to the voting polls. At 105 Fir Hill Street at the University Club next to the University of Akron's main campus, "a full house" of men gathered that evening for a special dinner meeting of the Akron Rubber Group. The rumors about a new and highly important development in rubber chemistry drew a large attendance. However, no one knew of the as-yet-undisclosed announcement of Duprene except Hezzelton E. Simmons, professor of chemistry at the university and secretary-treasurer of the Division of Rubber Chemistry, and, eventually, George Hinshaw, chairman of the group and future chairman of the division.

As was customary for meetings of the group, there was some accompanying entertainment. In this instance it was provided by a group of male singers called the Rubber Quartet, which included a Boston native and Goodyear's chief research chemist, Ray P. Dinsmore, chairman of the Division of Rubber Chemistry in 1927. (Dinsmore also was instrumental in the formation of the Akron Rubber Group.) At the meeting, a team of DuPont's top chemists and a young priest from the University of Notre Dame announced to the world the commercial availability of Duprene, the first general purpose synthetic rubber. Father Julius

Nieuwland, head of Notre Dame's chemistry department, whose early work in acetylene chemistry had pointed the way for DuPont's development of polymerized chloroprene, was one of the distinguished guests. DuPont was represented by Arnold Collins, Wallace Carothers, Ira Williams, F. B. Downing, W. S. Calcott, J. E. Kirby, and A. S. Carter.

Hinshaw, the organizer of the meeting, recalled that DuPont "made the announcement in what I thought was a very straightforward and truthful manner, explaining what its weaknesses were; why it cost so much; what it would do; what it would not do [in place of crude rubber]. . . . Consequently, we all left that meeting not feeling that crude rubber had been displaced from its throne, but that a material had been added to the rubber chemists' or rubber compounders' books that would enable him and his company to better serve the customers."

Also at that meeting was a young man who would later make a tremendous impact not only on the rubber industry but also on the chemical industry and his alma mater. Jim D'Ianni was a student of Hezzelton E. Simmons at the University of Akron. D'Ianni recalled, "My first exposure to synthetic rubber was, as an Akron U. student, attendance at that Akron Rubber Group meeting."[8] D'Ianni later, among many other great achievements, would play a major role in the U.S. government's Synthetic Rubber Program during World War II.

DuPont knew that its investment was worthwhile, but it took eight years before the company made a profit on Duprene. DuPont dropped its Duprene trademark on December 16, 1936, and referred to it as Neoprene. Bridgewater feared that the company would not be able to control the quality of the actual end-product that reached consumers. Under those circumstances, the more marketable term Neoprene was more appropriate to DuPont's role. However, along the lines of unsubstantiated folklore, DuPont's Oliver M. Hayden, in the Organic Chemical Department, said the reason for the name change was prompted by a complaint from a West Coast entertainer that DuPont was infringing on her stage name, Duprene. "Perhaps she thought DuPont would buy her off," Hayden speculated. By that time, however, DuPont had already withdrawn the trademark.

In 1934, a young, recent Ph.D. graduate from Ohio State University, Paul J. Flory, was assigned to Wallace Carothers's research group to assist in the product's refinement. By late 1939, and before the Louisville, Kentucky, plant was built, Bridgewater estimated that the company spent in excess of fifteen million dollars alone in Neoprene's development. During World War II, the government took control of the Louisville facility and spent an additional fifteen million dollars to expand its production capabilities.

Collins survived his hand-warming lab experiences of 1930 and other "accidents" and actually prospered. In 1981, he recalled some of those experiences with fond humor. "I've got a couple of scars here that I got by having something, a flask I was holding in my hand, blowing up," he said, laughing. "Oh, and then to show you how tricky it is, I always tell people about how I had made a sheet of plastic material, like a paper weight by polymerizing my compound. And one day I was showing it to Carothers [his boss, Wallace] and had it in my hand and said, 'See it's quite strong,' and I started to bend it and the thing blew up in my hand and in my face. Well, that was where we started."[9] The startling accident may have been a motivational push, but Collins's anecdote really understated how he and his colleagues at DuPont arrived at this new synthetic rubber. The reality is, the team indeed had some divine intervention.

Father Nieuwland was awarded (appropriately) the Priestley Medal of the American Chemical Society for his contributions to science. Like many good leaders, Bolton shunned the recognition limelight for his pioneering efforts in directing the team that developed Neoprene, the first successful and saleable general-purpose synthetic rubber in the world. He later guided a research effort that led to the discovery of nylon. He received many chemical industry honors for his achievements. For his contributions to the discovery of polychloroprene or Neoprene, Collins was awarded the Charles Goodyear Medal by the Rubber Division in 1973, the twenty-ninth recipient. Wallace Carothers was the least fortunate of the threesome.

A perfectionist in all his endeavors, Carothers was never satisfied with his own work. To help his parents and their financial woes, he bought them a house in Arden, Delaware, and moved in with them. He also suffered from bouts with depression, and the death of his favorite sister, Isabel, in January 1937 acerbated his own mental condition. Three months later, on April 29, 1937, in a hotel room in Philadelphia, he ingested a lethal dose of cyanide.

Collins, in his acceptance address for the Charles Goodyear Medal in 1973, reflected on his life and achievements:

> I have tried . . . to point out what were some of the factors in the experimental investigation which led to the discovery and final success of Neoprene. Some of these were imagination, which Nieuwland showed in hoping that cuprous chloride would activate acetylene, and which Bolton had in foreseeing that the further study of Nieuwland's divinyl acetylene and similar compounds would lead to a synthetic rubber.
>
> Another factor was Carothers' concern with thoroughness, something rare in intellectuals such as he was, which made him insist on having a pure divinyl acetylene to work with and thus led to the discovery of chloroprene. Some of the important decisions were just common sense, but since common sense is proverbially uncommon, such decisions are not often made.

DÉJÀ VU 8

We succeed only as we identify in life, or in war, or in anything else, a single overriding objective, and make all other considerations bend to that one objective.

GEN. DWIGHT DAVID EISENHOWER

History reflects that, strictly based on its boastful claim, World War I, the "war to end all wars," was a failure. However, the war provided a forum that nurtured and cultivated creative genius and experimentation, and the postwar years, from the 1920s to the 1930s, prolonged the trend establishing a solid precedent for American ingenuity in the twentieth century.

The 1920s have been called many things—the Jazz Age, the Age of Wonderful Nonsense, the Age of Intolerance, and, of course, the Roaring Twenties. Whatever the monikers, the decades and eras between the "wars," have never been called boring. Consider the following: Two monumental constitutional amendments were passed soon after the war's end—the Eighteenth, which made consumption and possession of alcohol illegal, and the Nineteenth, which gave women the right to vote.

For two decades following World War I, there was a roller-coaster ride of interest in science and technology, primarily dictated by the volatility of the economy. While the focus had changed from chemistry and chemical technology research for war causes to practical issues such as survival itself, synthetic rubber research maintained its subliminal presence in the minds of the rubber chemists. However, these ever-vigilant scientists would discover much too soon that their obsession with synthetic rubber—its availability, manufacture, and applications—would ultimately drive them back in a cruel twist of déjà vu to contributions to another global conflict. Most of the major powers of the twentieth century had experimented with synthetic rubber, but precedent would not give anyone a significant advantage when the world again would regress.

Hezzelton Erastus Simmons was born in 1885 at LeRoy, Ohio, in the northeast corner of the state in Lake County, just inland a few miles from Lake Erie. In 1913, after earning his undergraduate degree at Buchtel College in Akron and his master's degree from the University of Pennsylvania, he had returned to Akron to take over the chemistry department from Charles M. Knight, founder of the first courses in rubber chemistry in the United States.

"Hez," as he was called by his colleagues, was always looking for ways to promote the college, especially his chemistry students. One of the many opportunities for Simmons to spread Buchtel's rubber chemistry gospel came on April 7, 1919, at the meeting of the American Chemical Society that granted divisional status to the young professional society, the India Rubber Section. Simmons was asked to deliver a paper on physical testing of vulcanized rubber goods. In 1928, less than ten years later, Simmons would begin the first of his seven-year term as secretary of the Division of Rubber Chemistry. In 1931, he would become chairman of the

Akron Rubber Group and be involved in the worldwide announcement of the first successful commercial synthetic rubber, Duprene.

During his lifetime, Simmons was the author of many technical articles and associated with many scientific and fraternal groups. He made many contributions to rubber chemistry—including many papers in the division's primarily publication, *Rubber Chemistry and Technology*—and in 1952, he received the Charles Goodyear Medal from the division that he loved almost as much as he did teaching. He only absolved himself of full participation in the division in 1933 to become president of the University of Akron, a position he held until his retirement in 1951. During World War II, Simmons's input on synthetic rubber research would be valuable as he served as an advisor to the office of the rubber director on the War Production Board.

———————

Synthetic rubber originally was pursued as a substitute for natural rubber because of the decline in production from the plantations in Brazil and the universal fear of being cut off from the major supply. Its early development can be traced to the euphoria after Charles Goodyear's discovery of vulcanization in 1839 and Thomas Hancock's securing patent rights in England for Goodyear's finding. In 1879, thirty years after Goodyear's achievement, a French chemist, Gustave Bouchardat, further researched the 1860 findings of British chemist Charles Hanson Greville Williams.[1] Both men had worked with isoprene isolated from natural rubber. Williams had prepared his mixture in open air, but Bouchardat mixed his isoprene with hydrochloric acid, sealed it in a tube, and heated it for twenty hours.

Although the origin of each chemist's isoprene was natural rubber, Bouchardat claimed he had discovered *caoutchouc artificial*, or artificial rubber. But in 1882, Sir William A. Tilden, a British chemist, discovered he could make isoprene by heating turpentine. By 1884, he had refined his process, but not to the point where he could make a useful natural rubber substitute. Those efforts notwithstanding, research into a synthetic rubber was sparse for the next several decades, mainly because there was an abundance of natural rubber and the price was usually reasonable. No cost-effective substitute was yet in sight, nor was the demand for it. That would change.

Between 1914 and 1922, the price of natural rubber soared from 11.5 cents to $1.02 a pound.[2] Normally, a price increase of that magnitude would signal an increase in synthetic rubber research; however, World War I had forced chemists to focus their efforts on other activities—such as chemical warfare, bombs, and so on—and delayed their synthetic rubber research for a while.

By the 1920s and 1930s, most casual observers didn't realize it, but rubber usage was growing and rubber was becoming a very important material in the United States. America was consuming more than three-fourths of the world's rubber, and at that time, the majority of that rubber came from the plantations of Southeast Asia—British Malaya (Malaysia), Netherlands Indies (Indonesia), Ceylon (Sri Lanka), India, Burma, British North Borneo (state of Sabah, Eastern Malaysia), Sarawak, French Indochina (Vietnam), and Siam (Thailand)—where the equatorial climate favored the growth of *Hevea brasiliensis*, the world's finest transplanted rubber. For a variety of reasons, the wild rubber trees of the jungles of Brazil and South America were no longer the preferred source. The Brazilian trees were nearly tapped out and were producing less latex every year, and a leaf blight was slowing production even more. At the height of activity before 1940, about 97 percent of the world's rubber came from plantations in Asia.[3]

Natural rubber was a very important commodity, but it soon fell prey, like many things, to politics and became an instrument of leverage by large nations. In spite of the problems in South America, rubber usage continued to grow, but early on in the game, plans were being developed by the British, the owners of the majority of the plantations in Asia, to regulate production in an attempt to balance supply and demand for natural rubber. Great Britain's secretary of state for the colonies, Winston Churchill, organized a committee of inquiry, and, chaired by Sir James Stevenson, the Stevenson Plan materialized on November 1, 1922.

While there were many dissenters—primarily the Dutch East Indies, the United States, and the rest of the world—the Stevenson Plan still accomplished what the British wanted. In effect, it eventually raised the price of natural rubber to $1.21 a pound before it started falling three years later. However, thanks in part to U.S. Secretary of Commerce Herbert Hoover and inventor Thomas Edison, the British rescinded the plan on November 1, 1928, six years after its birth.

Natural rubber not only was a controversial political bouncing ball but also became a trustworthy economic indicator—an on/off button for research projects. It was a very common curve in the two decades between World War I and World War II. When the price of natural rubber was high, research activity of synthetic rubber also rose. Ironically, with demand for rubber products growing, it was unlikely that the price was going to drop dramatically. What was likely, and predictable, was the kindling of synthetic rubber research programs in the Soviet Union, Germany, and the United States between 1925 and 1932.

At the start of the 1930s, because of the Great Depression, the price of natural rubber was less than five cents a pound. By 1937, it was approaching twenty-five cents a pound—certainly not the high-water level of the early-to-mid 1920s but a trend significant enough to concern the industry, especially in the post-Depression era. This price spike meant the industry had to do one of two things: dramatically increase production efficiency to offset the escalating price of their most-consumed raw material or escalate the search for an alternative to that raw material—synthetic rubber. There wasn't much hope of improving manufacturing methods because, with a few exceptions, there were no major efforts being made in that direction.

Budgets were tight thanks to the lingering residual from the stock market crash of 1929. The only sane choice was to increase research in synthetic rubber development. Because of their unified cause and their voracious intent to establish self-sufficiency, Germany was the early leader in the area of synthetic rubber research, but it had established a precedent right after World War I. At that time, the Germans had learned from their experience with one of the first commercial synthetic rubbers, methyl. They used it when there was a blockade initiated, which prevented them from receiving natural rubber from the Asian plantations.

In the United States, rubber research that took advantage of those opportunities, not surprisingly, was led by four of the country's top rubber companies—the Firestone Tire and Rubber Company (which today comes under the large umbrella of Bridgestone Americas Holding, Inc.), the B. F. Goodrich Company, the Goodyear Tire & Rubber Company, and United States Rubber Company. The collective technical knowledge of key individuals in those companies was instrumental in the successful establishment of a synthetic rubber industry and eventually a united approach in those efforts.[4]

Ironically, two Russian scientists working for U.S. Rubber, Ivan Ostromislensky and Alexander Maximoff, initiated some of the first breakthrough synthetic rubber research efforts in 1922. In that year they made butadiene synthetically from ethyl alcohol and acetaldehyde, and in 1923 they produced synthetic rubber from butadiene by an emulsion polymerization process. They later received patents for those efforts as well as styrene.

The other rubber companies' achievements would follow in a few years, but the "Big Four" notwithstanding, contributions just as noteworthy also came from the periphery of the rubber industry. One of those "smaller" companies was Delaware-based DuPont, at that time officially called E. I. du Pont de Nemours and Company after its founder, Eleuthère Irénée du Pont (1771–1834). In 1925, a DuPont chemist, Elmer K. Bolton, who had spent time studying in Germany, conducted significant research on a natural rubber alternative. By 1931, Bolton and a team of chemists from DuPont joined by Albert M. Collins and Wallace H. Carothers, and a Holy Cross priest from the University of Notre Dame in South Bend, Indiana, Father Julius Nieuwland (see chapter 7), had developed Duprene, later called Neoprene. Primarily because of its versatility, Neoprene became the first widely used synthetic rubber compound. Later in the decade, a key member of that team, Carothers, would be credited with the discovery and development of nylon.

Another one of the early contributors to synthetic rubber research and development was Thiokol Chemical Corporation, ironically named after a product. Thiokol was discovered in 1926 and patented in 1927, the brainchild of two chemists, Joseph C. Patrick and Nathan Mnookin, who were just trying to invent a cheap antifreeze. While experimenting in their laboratory with sodium polysulfide and ethylene and propylene dichloride, they created a very smelly, gumlike substance. The gum clogged their lab sink, and there wasn't a solvent around that could remove it. Then it hit them—they had stumbled on to a synthetic rubber that had a resistance to solvents, oils, and certain gases. The name Thiokol was given to the substance from the Greek words for sulfur (*theion*) and glue (*kola*). Then, financed by a salt merchant named Bevis Longstreth, Thiokol Chemical Corporation was founded on December 5, 1929, in Kansas City, Missouri, but six years later the company moved to Trenton, New Jersey, because Kansas City residents complained about the plant's odor-producing fumes. Still, Thiokol was not the general-purpose synthetic rubber that the industry sought as a replacement for natural rubber.

Demopolis, Alabama, is located in the west central part of the state at the confluence of the Tombigbee and Black Warrior rivers, about halfway between two state capitals — Montgomery, Alabama, and Jackson, Mississippi. It was settled in 1817 by a few distraught exiles from Napoleon's court. Waldo L. Semon was born in Demopolis on September 10, 1898, and spent the first seven years of his one-hundred-year life in the small southern town. Young Semon was very inquisitive. "I became interested in chemistry while I was in grade school," he told Rubber Division historian Herbert A. Endres in 1966.[5] Semon's father, a civil engineer encouraged him in science, and Waldo "honored" him in those formative years by building a radio and then a canon. "My father attended Michigan Agricultural College [Michigan State University] in Lansing, Michigan, and was an avid student of chemistry. As a result, he always had plenty of books on chemistry in the house that sparked my interest." In 1905, the family moved to the Pacific Northwest, and Semon completed his education with a B.S. in 1920 and a Ph.D. in chemical engineering in 1923 at the University of Washington.

Semon was hired as a faculty instructor at his alma mater, but he soon realized it was difficult living on an instructor's salary while trying to support a wife and two children. He tried to supplement his teaching with consulting work, but the Washington legislature ruled that all faculty members had to donate consulting fees to the state. "I called Dr. [Harlan] Trumbull at B. F. Goodrich, who was my freshman chemistry professor and talked to him a while," Semon recalled. "He asked me if I was interested in a job at Goodrich in research. I said yes." Adios Washington, hello Ohio. "In 1926, I bundled my wife and kids into an old Ford, drove across the country, and on the first of June 1926 I reported to work at B. F. Goodrich."

It took Semon little time to establish his identity. His accomplishments would culminate in 116 patents, selection as the 1944 Charles Goodyear Medal recipient, and induction into the National Inventor's Hall of Fame in 1995 — among many other international achievements.

In spite of some of the pockets of excellence in synthetic rubber research scattered about the United States (with Akron as the magnetic center), the Germans were still the perceived world leaders, and they continued refinements of their earlier unsuccessful attempts from methyl during World War I. However, they were not only progressing technologically but also making some notorious geographical advances. Still, they needed rubber to get there.

Initially the Germans developed three types of methyl isoprene — "H" or *hart* (hard) rubber, "W" or *weich* (soft), and "B" for *balloon* (dirigible) coatings and wire insulation. The H rubber was used for submarine battery boxes and other electrical equipment and had more electrical resistance than natural rubber latex. The W rubber went into belts, hose, and tires. During World War I they produced about twenty-five hundred tons of methyl rubber. However, methyl was expensive, and it was not a satisfactory synthetic rubber. For example, cars that had tires with methyl W rubber had to be jacked up at night to prevent the tires from going flat.

With that knowledge, in 1926, when the price of natural rubber was making its climb, a German cartel, I. G. Farbenindustrie, A.G. (or I. G. Farben), picked up the synthetic rubber research baton that had been dormant while natural rubber was relatively inexpensive.[6] Realizing the ineffectiveness of methyl, I. G. Farben decided to focus on another substance, butadiene, with sodium as a catalyst. They gave this product the name Buna — a combination of the first two letters of butadiene and the first two letters of *natrium*, the Latin word for sodium.

From Buna's prototype, I. G. Farben developed a new copolymer, Buna S (styrene). However, the further refinement of the expensive Buna S was stopped by the Depression and, of course, the lower natural rubber prices. Also, Buna S was virtually impossible to process on existing machinery. However, buoyed by the announcement and success of DuPont's new Duprene, I. G. Farben revisited the styrene-based Buna

S formula and investigated the use of acrylonitrile. The result of their efforts with acrylonitrile was Buna N, and the new product offered something that Buna S didn't contain—oil-resistant properties similar to those of Duprene. Buna S still was a versatile general-purpose rubber, one that could be used as an alternative to natural rubber, so the Germans decided to commercialize both copolymers. Buna N was later changed to Perbunan and became a pioneer for the class of oil-resistant elastomers known today as nitrile rubbers. Buna N achieved small-scale industrial production and, for a while, enlisted the services of General Tire of Akron, Ohio. However, after a short testing period, and fearing that its machinery might be damaged from the testing, General gave the Germans a less-than-glowing report about the Buna N experience and then backed out.

While Buna N was pleasing to the Germans because it gave them a substitute for Duprene and Thiokol, it was not suitable for tires, and the ripples of governmental intervention were turning into tidal waves. However, the prevailing attitude among the industry was reflected by B. F. Goodrich's Harlan Trumbull, Rubber Division chairman in 1937. Slightly miffed by the attitude of his colleague William C. Geer toward synthetic rubber, he said, "It was a reflection of the top management's belief in 1919 that if the Germans could not succeed in making synthetic rubber, nobody could do it."[7]

There was a mammoth political machine building in Deutschland, and the National Socialist Party, the Nazis, were crafting their empire. They wanted total control of Germany and this included self-sufficiency with regard to, among many other things, rubber. In 1936, the Nazis selected Hermann Wilhelm Göring to oversee the country's "Four-Year Plan." The plan was a series of economic reforms designed to develop public works projects and increase automobile production (including developing the autobahn system), other building and architectural projects, and synthetic rubber production. This was a somewhat covert attempt to mask the buildup of their military forces, a bold violation of the terms set up by the Allies of World War I in the Treaty of Versailles.

In 1937, partly in reaction to their unsuccessful attempts to obtain a license from DuPont for Duprene, the Germans revived their interest in

Buna S. They realized that unlike Buna N, Buna S could be mixed with natural rubber. So they began manufacturing different varieties of Buna S at plants in Hüls and Schkopau,[8] and when they compounded it with carbon black, Buna S was significantly more durable than natural rubber—excellent for production of tires. The plants at Hüls and Schkopau were very efficient. They began producing about two hundred tons a month, and in their first year they reached a capacity of seventy thousand tons annually. In effect, this achievement signaled the start of industrial production of the first general purpose synthetic rubber. Today, we call it styrene-butadiene rubber, or SBR.[9]

As a corollary to the resurgence of Buna S, Friedrich Bergius, a chemist with Badische Anilin und Soda Fabrik (BASF), developed a process to convert brown coal into oil. Since Germany had no oil fields but plenty of coal, this was a major development for the country. However, the Bergius process was not cost-effective and couldn't produce quantities suitable for the new political movement. The company needed some manufacturing expertise and looked toward the United States.[10] Early in 1925, BASF directors, visited, among several companies, Standard Oil of New Jersey and its crafty chief executive Frank A. Howard. Standard and BASF developed a partnership that purchased Gasolin, A.G., a company that produced and distributed gasoline in Germany. To protect his company's interest, Standard's Howard visited BASF's Ludwigshafen, Germany, facility. Howard was impressed with the capabilities and in the potential for a partnership with BASF to help further develop the Bergius process. However, since BASF was a part of the I. G. Farben conglomerate, they needed approval from their entire group for such a venture. In 1927, with the protocol now well defined, Standard signed an agreement with I. G. Farbenindustrie to offer technical expertise in the development of the Bergius Process in the U.S. with I. G. Farben retaining patent and licensing rights, but Standard could share in half the royalties. In October 1930, the two companies formed JASCO, an acronym for Joint American Study Company, which in essence gave Standard access to patents for Buna rubber, owned by I. G. Farben.

However, when war grew imminent, the companies signed the Hague Agreement, a covenant of the World Intellectual Property Organization (WIPO) concerning the international deposit of industrial designs. The pact was originally created on November 6, 1925, in London. The agreement enabled Standard to retain all the rights of the JASCO process, including synthetic rubber (Buna) in the United States, Great Britain, and France while I. G. Farben had all the rights outside of those countries. The oversimplified reality was that Standard kept the rights to German synthetic rubber technology.

While Germany was involved with its own visions of grandeur, there was plenty of activity in East Asia, primarily initiated by the Japanese Empire. As early as 1934, Japan had divulged to the rest of the world its intent to expand its own meager geographic boundaries. That's when they invaded northern China, specifically, Manchukuo; established a puppet state; and crowned Pu Yi as its emperor. Manchukuo today is the city of Xinjing, located near Shanghai.

Three years later, the Japanese signaled their official alienation from the United States, Great Britain, France, and Italy by withdrawing its support of the Washington Naval Treaty signed by those countries in 1922, which limited the size of its navy.

During the mid-1930s, when all the rubber companies were looking for a multipurpose synthetic rubber, two U.S. companies took the lead in development—B. F. Goodrich and Waldo Semon and Goodyear and Loren Sebrell. Both men visited Germany in 1936 and 1937 to investigate Buna S, the recognized premium synthetic rubber of the time. "I remember driving around in a car with—what was obvious to me—tires that were made from Buna S," Semon said. "But it was impossible to get any technical information on synthetic rubber from the Germans. We had been working in our own labs quite successfully in making polymers, so I came back more enthused than ever, knowing we could develop a method of our own."

Semon had previous success in converting polyvinyl chloride from a hard, unworkable substance to a flexible one. He initiated the age of vinyl, today an industry worth more than twenty billion dollars. The synthetic rubber needed for tires took additional research help, but Semon had plenty of that with his former professor Trumbull, David Craig, Ben Garvey, Ed Newton, Ernest Bridgewater, Art Juve, and

Harold Gray—also all prominent members of the Rubber Division of the American Chemical Society. Six of the seven Semon "associates" would eventually serve as a chairman of the Rubber Division, and the other, Craig, would become editor of *Rubber Chemistry and Technology*.

In 1937, as B. F. Goodrich continued its work, William J. Sparks and Robert M. Thomas of Standard Oil of New Jersey developed a synthetic rubber called butyl whose primary characteristic was its impermeability to air, a major requirement for tire inner tubes. In two years Goodrich would have a new, very progressive, and forward-thinking president, John L. Collyer, former managing director of the massive Dunlop Rubber Company in London. While in England, Collyer attempted to alert the British government about the ulterior motives of Japan and Germany, but he was ignored. Collyer essentially told his new Goodrich researchers to develop a product, and then he formed a joint venture with Phillips Petroleum Company called Hycar Chemical and appointed Semon as a vice president.

As the synthetic rubber train gained momentum, the Rubber Division realized increasing opportunities to simultaneously climb on board and stoke the fire. In 1938, at the invitation of the Institution of the Rubber Industry of Great Britain, twenty-two members of the Rubber Division attended the first International Rubber Technology Conference in London. Rubber Division chairman Archie R. Kemp of Bell Telephone Laboratories led the delegation. Some of the others representing the division included Sam Gehman and Ray P. Dinsmore of Goodyear, L. E. Oneacre of Firestone, A. A. Somerville of R. T. Vanderbilt, William B. Weigand of Columbian Carbon, J. T. Blake of Simplex Wire and Cable, Norman Bekkedahl of the National Bureau of Standards, W. F. Busse of B. F. Goodrich, Ira Williams of DuPont, and Howard I. Cramer, G. Stafford Whitby, and Hezzelton E. Simmons of the University of Akron.

The three-day event included 105 papers from eighteen countries. Papers were delivered in the following categories: applications of rubber, durability of rubber, compounding materials, physics of rubber, chemistry of rubber, latex, general technology, plantation subjects, and synthetic rubber-like material. Royce J. Noble, who was made a Fellow of the Institution of the Rubber Industry, commented, "The success of the undertaking was due in large part to the complete cooperation of leading technical organizations throughout the world including the American Chemical Society's Rubber Division."[11]

I. G. Farben, the German conglomerate, was represented by four of its top chemists and researchers and dominated the papers on synthetic rubber-like materials. The following spring, in April 1939, likely intrigued by presentations from the I. G. Farben entourage, the Rubber Division decided to devote its meeting in Baltimore to synthetic rubber and elastic polymers. To add international credibility to the program, the division invited a guest speaker from I. G. Farben to present a scientific paper on Buna rubber. The speaker was Albert Koch, who intrigued the audience with his thoughts on how "to stimulate interest in synthetic rubber in the United States."[12]

On September 1, 1939, Germany invaded Poland on the ridiculously false pretense that some Polish soldiers had attacked a German radio station at Gliewitz on the German-Polish border. The "soldiers" actually were concentration camp inmates forced by the German SS to dress like Polish soldiers and stage the chicanery.

With Germany already at war with Europe and in virtual control of the synthetic rubber industry, and the expectation that Japan would escalate its efforts in Asia growing steadily, President Franklin D. Roosevelt declared rubber to be a "strategic and crucial material" for the nation's existence. As a result, the Rubber Reserve Company (RRC) was formed in June 1940 to stockpile natural rubber, and while the RRC was materializing, the Advisory Committee of the Council for National Defense was established in the same month with Clarence Francis, president of General Foods, as its chairman. Known as the Francis committee, it had a short life. An inherent redundancy between the committees was eliminated when the RRC took over the functions of the committee in October 1940.

On December 7, 1941, the United States' trepidations materialized as Japan attacked the U.S. Naval Base at Pearl Harbor, Hawaii. The next day, while President Roosevelt asked Congress for a declaration of war, the Japanese also invaded Malay (Malaysia) and the Dutch East Indies (Indonesia), cutting off 90 percent of the world's valuable supply of natural rubber. Those days continue to haunt America.

SUMMA CUM LAUDE

9

Life should not be estimated exclusively by the standard of dollars and cents. I am not disposed to complain that I have planted and others have gathered the fruits. A man has cause for regret only when he sows and no one reaps.

CHARLES GOODYEAR

No one individual embodies the spirit of the Rubber Division of the American Chemical Society more than Charles Goodyear, the world's recognized inventor of vulcanized rubber. He was a pioneer, an entrepreneur, and just plain loved to experiment with things. To say that Goodyear was obsessed with rubber in the 1830s and throughout his life is a classic understatement multiplied many times. It's like

80

saying bees love honey. Goodyear swarmed over rubber. If he could have, he would have eaten rubber—maybe he did because in his later years he suffered from dyspepsia (acid indigestion) and gout. Goodyear operated in the same manner as the sailor who was commanded by his chief petty officer to paint anything that didn't move. So Goodyear would "paint" almost anything with rubber—banknotes, musical instruments, jewelry, and, yes, even ships.

This spirit, often called insanity by his family and friends, engulfed him the rest of his life. In 1839, however, with more than half of his years on earth almost completed, Goodyear's determination and passion for rubber finally gave him what he craved—satisfaction for just having found something worthwhile, nothing more than that. In his home of Woburn, Massachusetts, on the periphery of Boston, he discovered vulcanized rubber.

Technically, Charles Goodyear's discovery was empirical, and it took many years before the process he had used was defined scientifically. That didn't bother him. Goodyear never reaped the financial benefits of his discovery, and he didn't even create the tire company in Akron, Ohio, that was named for him. He died in 1860 at age fifty-nine, $200,000 in debt, but his spirit and his pursuit of invention and discovery for their own sake are the foundation of the motivation behind the Rubber Division's premier honor, the Charles Goodyear Medal.

The gilded Grand Ballroom at the Copley Plaza Hotel in Boston was filled to its one-thousand-person capacity on the evening of September 13, 1939. The VIP speakers included Paul W. Litchfield, president of the Goodyear Tire & Rubber Company; Karl Taylor Compton, president of the Massachusetts Institute of Technology; and James Bryant Conant, president of Harvard University. Behind the speaker's table in the ornate room hung a large original painting of Charles Goodyear, befittingly on hard rubber and draped by red-white-and-blue bunting. The event was the main banquet at the joint general meeting of the American Chemical Society and the Division of Rubber Chemistry orchestrated to honor the centennial of the discovery of vulcanization in nearby Woburn and, specifically, its former inhabitant, Charles Goodyear. The celebration was more than an overnight success.

The meeting, like many conducted during the division's formative decades, was also the launching mechanism for greater things. The idea for the centennial celebration materialized in a Division of Rubber Chemistry meeting in 1937 in Rochester, New York. Division chairman Harlan L. Trumbull of B. F. Goodrich said that he had been approached by "certain prominent chemists" within the organization to honor Charles Goodyear on the one-hundredth anniversary of his discovery. At the 1939 spring meeting in Baltimore, the honor of selecting a committee to organize such an appropriate event went to Rubber Division chairman George K. Hinshaw, appropriately an employee of the Goodyear Tire & Rubber Company. "I selected a group of men from the industry from all across the country to serve on the committee to organize the event," said Hinshaw, who also was chairman of the meeting committee. The committee had the additional responsibility of soliciting funding. "When each of the members came back from approaching their organizations and companies," Hinshaw said, "we had more than enough money to put on a very successful meeting."[1] It also helped that Hinshaw was a former chairman of the Akron Rubber Group, which at the request of Hezzelton Simmons at the University of Akron organized the meeting in 1931, when DuPont announced Duprene to the world.

Hinshaw had become chairman of the division in a fashion similar to his selection by Simmons to organize the DuPont meeting just seven years prior to that. "I was sucked in," he said with a laugh. "Sid Caldwell [division chairman in 1935] came to me and asked for suggestions on the qualifications of someone who might help organize this Charles Goodyear celebration. I told him not to mess around and just pick the best man for the job. Well, a few days later, I found out that all I did, in talking to Sid, was talk myself into a job!"

Chairing the local arrangements committee for the event was John M. Bierer, division chair in 1926 and chairman of the Boston Rubber Group in 1928. Bierer selected C. R. Boggs, A. A. Somerville, and W. B. Weigand to assist him. "We had seventy-two companies donate approximately sixteen thousand dollars for the event," Bierer told division historian Herbert A. Endres in 1966. "We also gave each guest a copy [reproduction] of the original books written by Charles Goodyear and Thomas Hancock [that recounted their discoveries]; a sterling silver cocktail shaker and various other souvenirs."[2] In addition, Bierer said that the speeches given by Litchfield, Conant, and Compton were "broadcast on radio from coast to coast."

Trumbull had only one other pertinent comment about Hinshaw's organization of the meeting. In referring to Goodyear's selection of sulfur as the key ingredient in vulcanization, he said, "Hinshaw opened this noteworthy meeting with a prayer, and any diabolical curse pertaining to [Goodyear's] choice of the abyssal brimstone was thereby exorcised officially." Hinshaw dismissed the reference and even his "selection" as committee chair and added, "In the end I was very grateful. It was an event that I've been very proud of all of my life."

The Boston meeting not only produced the centennial celebration for Charles Goodyear but also was the birth of the Charles Goodyear Medal, annually the top honor bestowed on an individual for his or her "contributions to the science or technology of rubber or related subjects."[3] It was the extension of a proposal made at the spring meeting in Baltimore earlier that year—the establishment of the Charles Goodyear Lecture. In an executive committee meeting, the official rules and regulations were adopted as "founded by the Division of Rubber Chemistry of the American Chemical Society in Commemoration of the One Hundredth Anniversary of the Discovery of the Vulcanization of Rubber by Charles Goodyear."

The governing document also provided for the creation of the Lecture Committee with the division's immediate past chairman in charge of the group. The members of the committee were limited to seven past chairmen of the division who were "active members." Like the U.S. Constitution, the original regulations or criteria for selection were general enough to provide the committee with sufficient latitude to nominate the appropriate individual. The essential guidelines remain the same today: "To recognize those who have made outstanding contributions to the science of rubber or related subjects."

The committee was commissioned with selecting one living individual a year who met the criteria. In return, the award winner only had to show up and deliver a lecture/paper. For those efforts, he would get a "suitably inscribed certificate stating that he was the Charles Goodyear Lecturer," and ten days prior to delivering the lecture he would receive an honorarium of two hundred dollars. The emphasis was made on giving the recipient the money before he traveled, assuming that it would help him pay car, boat, train, or air transportation to the meeting site. The actual lecture "ceremony" would be conducted "at one of its regular meetings," with the season to be determined.

The selection wheels for the first recipient weren't set in motion until the spring meeting of 1941 in St. Louis. In the business meeting held at the end of the day on Friday, April 11, lecture committee chairman E. B. Curtis announced the selection of David Spence of B. F. Goodrich as the Charles Goodyear Lecturer for 1941. Curtis also reported that Spence would deliver his lecture during the annual meeting of the division in September at Atlantic City, New Jersey.

Spence was a logical choice. Born in 1881 in Udny, Aberdeenshire, Scotland, approximately six hundred miles north of London and just inland from the coast of the North Sea, he received his undergraduate degree in chemistry at Royal Technical College in Glasgow in 1903 and his doctorate from Jena University in the state of Thuringia, Germany. Spence was an assistant professor of biochemistry at the University of Liverpool from 1906 to 1908, when he left for a job in Akron, Ohio, and Diamond Rubber Company as director of the research laboratories. When Diamond was absorbed by B. F. Goodrich in 1912, Spence became a part of the prominent BFG fraternity which included William C. Geer, Arthur H. Marks, George Oenslager, and Walter W. Evans. In 1925, Spence just missed the arrival of Waldo Semon and left B. F. Goodrich to become vice president of research for Intercontinental Rubber Company, headquartered in New York but with operations in Salinas, California. Years later, Intercontinental would be the unlikely acquisition of a company called Texas Instruments.

In the late 1920s, Intercontinental Rubber became heavily involved in guayule, a bush that grows well in dry, arid areas and produces a rubber that, when combined with reclaimed rubber, has limited applications. Intercontinental wanted Spence to investigate improving guayule's yield. In his research, Spence published a paper outlining how the United States could not become dependent on foreign supplies. The War Department read it and sent two high-ranking officers (one was Dwight D. Eisenhower) to investigate the Salinas operation. During

World War II, the Department of Agriculture approved a twenty-five-million-dollar project to plant forty-five acres of guayule in the Southwest.[4]

In 1936, Spence worked in California at Stanford University's Hopkins Marine Station, where he confined his research primarily to extraction of rubber from guayule plants. Spence was assisted for one year at Hopkins by a recent Ph.D. graduate from Stanford University, John Douglass Ferry. Ferry would win the Charles Goodyear Medal in 1981 representing the University of Wisconsin.

On September 11, as misfortune would manifest itself as a common occurrence in 1941, Spence, the first recipient for what would evolve into the division's top award, was a no-show for the inaugural Charles Goodyear Lecture in Atlantic City. Division chairman R. H. Gerke embarrassingly announced that Spence "was unable to deliver the lecture because of illness." Spence did not deliver the Charles Goodyear address and did not collect two hundred dollars. However, the following resolution appeared in the official minutes: "Be it resolved that David Spence has been elected officially the first Charles Goodyear Lecturer; that this resolution be spread on the published records of the 1941 meeting of the Division of Rubber Chemistry; and that a suitable scroll certifying this honor be presented to him by the Charles Goodyear Lecture Committee." At the same meeting, the committee revised the rules to accommodate absentees. It was a fortuitous decision.

On September 11, 1942, in Buffalo, New York, "in the interest of wartime secrecy," Rubber Division chairman John N. Street of Firestone announced that the presentation of the second Charles Goodyear Lecturer, Lorin B. Sebrell of the Goodyear Tire & Rubber Company, scheduled for Memphis in the fall, was cancelled. In addition, the scheduled papers of Walter W. Evans and John M. Bierer were withdrawn by "censors."[5] Street also reported that because Sebrell's rescheduled 1943 lecture was not yet cleared by Washington, it didn't allow the committee the proper amount of time to select an appropriate lecturer for the following year (1944). In Detroit on April 15, 1943, Sebrell finally received the opportunity to give his lecture. Introduced by Archie R. Kemp of Bell Telephone Laboratories, Sebrell delivered a speech titled "The Second Mile." It was the first official lecture delivered in the history of the award.

The war, which occupied most individuals' minds those days, continued to play havoc with the Charles Goodyear Lecture. In the fall of 1943, at a New York City meeting of the executive committee, chairman-elect John T. Blake, at the request of some lecture committee members, asked Kemp to sit in on a review of the award's criteria. There were still issues that needed clarification. The committee also asked if Blake would extend an invitation for Spence, the first recipient, to deliver his lecture in 1944.

Blake and the committee, though, would have to wait again. World War II forced the cancellation of another Charles Goodyear Lecture. However, Blake was not going to let cancelled meetings prevent another individual from being honored. At the April 26, 1944, meeting of the executive committee, Blake sent out a letter requesting nominations for the honor. In the first ballot conducted by mail, Waldo N. Semon of B. F. Goodrich was selected as the 1944 Charles Goodyear Lecturer. True to form, the fall 1944 meeting was cancelled, and for the first time in its history, the division's general meetings for the year (1945) were cancelled. Information was exchanged, decisions were made, and meetings were held only by mail.

The year 1946 was a glorious one for everyone in the world, and the Division of Rubber Chemistry of the American Chemical Society joined in the celebration. After virtually two years without in-person meetings, on April 10, 1946, the division conducted the 109th meeting in Atlantic City, New Jersey. Semon, only the third recipient of the division's top award (and at the time, the only one), finally delivered the lecture for his 1944 honor. However, as the result of a letter ballot prior to the meeting, the name of the Charles Goodyear Lecture would be changed to the Charles Goodyear Medal with the 1946 recipient.

Willis A. Gibbons was chairman of the Division of Rubber Chemistry at the time and knew the situation as well as anyone. He had been chairman for the last two years—the last one to hold that distinction since 1919—because of the war. Prior to Gibbons, Lothar E. Weber actually served three terms from 1916 to 1919. The first to have that honor of consecutive terms was the pioneer chairman, Charles Cross Goodrich in 1909, the year the India Rubber Section was founded. Goodrich held the position for three years and was followed by David A. Cutler of the Alfred

Hale Rubber Company and then Weber until 1919, when Firestone's John B. Tuttle initiated the standard practice of one-year terms.

Gibbons revealed that the committee wasn't pleased with the lecture structure. "This really hadn't operated too well," he admitted. "It seemed to us that if the idea was to recognize merit, it would be far better and simpler to have a medal rather than the foregoing arrangement. So the committee voted to have a medal to be awarded every year."[6] The bylaws were then revised to incorporate most of the changes that exist today. The honor of being the first official Charles Goodyear Medal winner went to Ira Williams (1946) of DuPont, who delivered an appropriately themed lecture titled "Vulcanization of Rubber with Sulfur" on September 13, 1946.

The residual effects of the war were still having an impact on the division's scheduling of meetings as the "director's meeting" on May 27, 1947, prompted the directors vote to cancel the fall meeting. However, they agreed to have a meeting in the spring of 1948, and Chairman Walter W. Vogt of Goodyear proudly announced that a design for the new Charles Goodyear Medal was in the development stage. In the September 17, 1947, meeting in Chicago, Vogt showed samples of the distinctive medal and announced that George Oenslager of B. F. Goodrich was selected as the 1947 recipient of the Charles Goodyear Medal and that Oenslager would become the first individual to receive the actual medal. Vogt quickly added, "Similar medals are to be presented to Messers Sebrell, Semon and Williams." Likely in a more conciliatory tone, he announced that the budget would be increased by sixteen hundred dollars to handle the expenses of procuring the medals.

On April 21, 1948, at the Hotel Sherman in Chicago, Oenslager delivered his Charles Goodyear Lecture. However, the only mention of the event in the official minutes of the meeting was a note, hand-written in pencil in the margin, that read, "Geo. Oenslager was given medal."[7] The item that preceded it documented the executive committee's vote to present a "scroll and a medal" to David Spence, the first honoree, at the summer meeting in Los Angeles.

On July 23, 1948, seven-plus years after his selection, Spence received the Charles Goodyear Medal from Harry E. Outcault of St. Joseph Lead.

In the years that have followed, the format and criteria have remained essentially the same. The initial two-hundred-dollar honorarium is now (2009) six thousand dollars, primarily to cover the costs of transportation, hotel, meals, and so on for the honoree and a guest. In 1958, Raymond F. Dunbrook of Firestone thought it would be prudent to take advantage of a sale on scrolls, so he purchased a fifty-year supply at the great price of two dollars each. Prior to the existence of today's all-inclusive Awards Committee, the opportunistic Ralph Graff had similar foresight in the mid-1970s and purchased five gold-plated medals at one time "at a really great price considering the cost of gold these days," he said.[8]

In 1958, Harlan L. Trumbull, division chairman in 1938, wrote a four-page procedural review of the medal for the division's records. The contents of his candid appraisal are noteworthy only to a select few, but his efforts in the thoroughly investigated project demonstrate the dedication of the true volunteer spirit that permeates the organization. In 1978, Tom H. Rogers of Goodyear—division chairman in 1969—authored a historical review of the medal with a solid quantitative analysis. Rogers' paper was published in *Rubber Chemistry and Technology*. His excusable left-brain (statistical, logical, objective) examination is interesting. However, times indeed have changed.

Today, it would be unfair to the spirit of the award and to former recipients to categorize their efforts on the basis of their affiliation with companies, universities, and organizations or the subjects of their papers. With all due honor and respect to each of those great individuals, Paul J. Flory, who won a Nobel Prize in chemistry in 1974, still stands alone among the Charles Goodyear Medal winners for his pioneering work in the molecular interpretation of rubber elasticity. There are few who would argue that.

One might also reflect that Jim D'Ianni in 1977 was the last division chairman to have received the medal. To critics and supporters alike, the trends of the selection of tomorrow's honorees aren't like grabbing feathers in the wind. Rather, they are suggestive of an adaptation to a constantly changing, better-educated, and faster-growing universe focused on technology, science, and its many applications. Elastomers will always play a role in those trends. The common thread that continues to pass through the ribbon of the medal itself is the pioneering spirit of Charles Goodyear himself.

It's also the spirit of Loren B. Sebrell of Goodyear, the second medal recipient, who was born and raised in Alliance, Ohio. After graduation from Mount Union College in 1916, he took his B.S. in chemistry and entered the Chemical Warfare Service during World War I. He was inspired by Roy Bowman, his high school teacher. After he retired from the International Latex Corporation in 1959, Sebrell was elected to the Delaware House of Representatives in 1966, serving two terms.

It's the spirit reflected in J. Reid Shelton, the 1983 recipient who was raised on a farm in Iowa and, like so many successful people in life, took the advice of a high school teacher and entered an essay contest. The one-room school had no chemistry class, but English teacher Claudia Richards encouraged Shelton to write about something he loved. He wrote about chemistry in agriculture and the industrial use of corncobs. It was his own personal seed of knowledge as he left Clio, Iowa, to receive his B.S., M.S. and Ph.D. Ultimately, Shelton retired in 1977 as professor emeritus of Case Western Reserve University in Cleveland. He was just an enthusiastic, inquisitive farm boy.

It's the spirit of D'Ianni, the 1977 recipient and son of an Italian immigrant who grew up in Akron — North Hill — and was a star student at Bryan Grade School, just a block from where his parents lived. He spoke fondly of Mary Fitzgerald, his high school teacher. "She was my favorite," he said. "Because of her, I jumped grades four times and finished grammar school in six years." D'Ianni later would add others to his list of favorite teachers who helped him along the educational pathway. He received almost every honor a rubber chemist could receive in life, but as much as he received, he gave twofold in return. A defining act of his life came after his highly successful career at Goodyear and his many contributions to the Synthetic Rubber Program during World War II. In August 1999, D'Ianni donated $1.75 million to his alma mater, the University of Akron.

It's the spirit reflected in the words of Goodyear's Tom H. Rogers, who wrote a brief history of the award for the division in *Rubber Chemistry and Technology*:[9] "They certainly all possess the ability to pursue aggressively their experimentation and to sell the resultant product to their management. The large numbers of their patents show that they all had the ability to carve a niche out of the existing art, whether it [was] large or small, for themselves. No two used the same technique, but they all had a flair for public relations in convincing their contemporaries that they were the true leaders in their fields."[10]

It is the spirit reflective of everyone who has received the Charles Goodyear Medal—and who will ever wear the medallion. It is awarded to those individuals not just for what they discovered, but just as much for how they got there. Reflective of the true spirit of Charles Goodyear, it is their will to continue research, in spite of adversity or other obstacles, not only for their own benefit but also for the benefit of future generations.

CRITICAL COMMODITY

Knowledge is power.

SIR FRANCIS BACON

War is a great case study on paradoxes. It brings out both the worst and the best in human nature. When the Japanese attacked Pearl Harbor, Hawaii, on Sunday morning, December 7, 1941, the world was perched on the branch of yet another planet-wide conflict, one that would kill and cripple more people and destroy more land and buildings than any war in history. Yet it would inspire many innovations that would serve as catalysts generating thousands of products to help rebuild the world as no one could have imagined. It would unite, among others, Americans— regions, states, cities, townships, churches, schools, and organizations—with such a poignancy that the memories and residual effects would permeate and remain with society even today. Many would bond for a common cause that they had never experienced before. World War II was the second major global conflict of the century, and, in an ironic twist of events, it would simultaneously divide countries while unifying millions.

One of the key and often underpublicized contributions to the United States' success in World War II was the massive effort of the Synthetic Rubber Program. While the military was conducting its battles in the Atlantic, Pacific, Europe, Africa, and other strategic locations throughout the world, most of the equipment they used required rubber. Whether in the air, on land, or at sea, the vehicles, equipment, and personnel used many things that contained rubber.

A report issued early in the war effort revealed some poignant examples of the military's urgent need for rubber. It said that a U.S. Navy battleship required more than seventy-five tons of rubber—for life rafts, ponchos, boots, shoes, millions of feet of hose, and thousands of miles of rubberized signal wires. The B-17 Flying Fortress, a staple of the U.S. Army Air Corps, used more than half a ton of rubber, while the standard ground combat vehicle, the Patton tank, used a ton of rubber. In addition, a soldier used a mind-boggling 194 pounds of rubber, compared with only 32 pounds used by his World War I counterpart.[1] Rubber was indeed a critical commodity to the war effort, but knowledge was going to help overcome any obstacle.

In 1940, there were a few developing concerns. Because it was readily available and relatively inexpensive, natural rubber was the commodity of choice. It had been that way for decades. Only a small percentage of the rubber used—less than 1 percent—was synthetic. With the advances of the Japanese military toward the plantations of Southeast Asia, the source of the majority of the world's natural rubber was in jeopardy. Fears became reality when the Japanese bombed Pearl Harbor and almost simultaneously invaded the countries that produced 90 percent of the world's natural rubber supply.

With the woefully weak supply of natural rubber available in inventory to the United States for its war efforts (an estimated 570,000 tons),[2] America had to rapidly find an alternative source of rubber or face the possibility of losing the war. Through the contributions of thousands of individuals and organizations, the rapid creation of the Synthetic Rubber Program helped the United States win the war. It was a monumental achievement, and the rubber industry rose as one collective entity to perform as the major donor to that endeavor. "The people in the rubber industry saw the emergency coming long before the people in the government saw it," said Benjamin Kastein, longtime historian of the Rubber Division of the American Chemical Society. "Natural rubber was selling for five cents a pound, and everybody realized that synthetic rubber was going to cost around thirty cents a pound, and you can't compete with thirty-cent rubber against five-cent rubber on a private company basis. So it had to be a government-funded project. And so they were looking to the government to step in and set up a program."[3]

In 1940, with his B.S. in chemistry from the University of Wisconsin, Kastein joined the Firestone Tire and Rubber Company in Akron, Ohio. A year later he would join the Akron Rubber Group and the Division of Rubber Chemistry of the ACS. Kastein was impressed with the division's structure. "They were all about relationships," he said. "Information. We'd get probably more information in the hallways than in sessions, and we'd share information with our counterparts at the other rubber companies."

The other place where the U.S. rubber industry received a great majority of its information was through the standard, well-respected, and most prolific medium of the first half of the century—the scholarly journal. In this case, *Rubber Chemistry and Technology*.[4] In order to receive industry-specific and technical information marketed solely to the rubber chemist, *RC&T* was "the Bible." The major commercial publications, such as *India Rubber World* and *Rubber Age*, provided useful information, but when a researcher wanted specific data on a specific rubber-related subject, *RC&T* filled the void. In particular, prior to World War II, *RC&T* kept Rubber Division members as well informed as any other medium in the world. The papers published in it came not only from the United States but also from all over the world. The list of international contributors to its pages is lengthy and impressive.

Right up to the point when the United States entered the war in December 1941, *RC&T* received regular contributions from the Soviet Rubber Industry, State University of Moscow, University of Leningrad,

Molotov State University, Netherlands Government Rubber Institute, University of Vienna, Indian Institute of Science, University of Geneva, Tokyo Technical Institute, Yokohama University, Kaiser Wilhelm Institute, and a very familiar name in the rubber industry, Germany's I. G Farben. And there were many others.

I .G. Farben began contributing technical papers as early as 1931 with Dr. G. Susich's "Fusion Curve of Natural Rubber." In 1937, I. G. Farben's Albert Koch submitted a paper titled "Specific Properties of Buna Rubber." This particular paper earned Koch an invitation by the Rubber Division to speak personally at their 1939 fall meeting in Baltimore (see chapter 9).

Kastein was exactly right when he said that the rubber industry knew what needed to be done even before the war had started. Because of its mantra of sharing information—primarily through *RC&T*—members of the division were armed with valuable technical knowledge as they routinely left their own companies to contribute as only they knew how, by using their minds to help develop a program to manufacture synthetic rubber. In a manner similar to that of a soldier, these knights of the rubber industry dropped everything to help the war effort.

The information that Rubber Division members in particular had gained via *RC&T* and their meetings prior to World War II helped them in their many individual contributions to the development and ultimate success of the Synthetic Rubber Program. Kastein estimated that every member of the Rubber Division contributed time and expertise to the program's cause. While the division as an organization did not participate in the Synthetic Rubber Program, its officers and members joined in a massive, largely volunteer information-sharing effort that in the end proved critical to the program's success.

The members of the Rubber Division—with its hundreds of volunteers who comprised the organization—played a role in the war's outcome. These members—chemists, engineers, researchers, and scientists—came from all of the major rubber companies, their vendors, or suppliers and academia, and they used the Rubber Division to share information, data, and thoughts throughout the war, like families discussing a variety of factual and emotional issues. The division offered them a forum to freely provide objectivity to the nation's common cause célèbre. The ever-present common thread for most of the participants, besides rubber, was membership in the Rubber Division of the American Chemical Society. The division wasn't the *industry* or the *Synthetic Rubber Program*, but as Kastein put it, "The Rubber Division was kind of a *mirror* of what was going on in the rubber industry."

The many roles taken by division members during that tumultuous era began rather placidly but evolved collectively into one of the major contributions of the war effort—and of the rubber industry, the industries that relied on rubber and, in reality, the world. Even the Rubber Division's general meeting in the spring of 1940 in Atlantic City, New Jersey, had a patriotic theme: "Rubber for Defense." It was the development of not just a new product or two but also of an entire industry—a synthetic rubber industry.

Early on in the development process, when the industry was dealing with government entities, they would give this new general-purpose synthetic rubber an appropriate military-type name. They called it GR-S, a catchy military-type acronym for government rubber–synthetic.

The task was awesome and unparalleled. The planet was involved in war and needed help from every human being, every company, every organization, every politician, and every government entity to stop the aggression. While the tripping mechanism that motivated the participants was like a cascading set of dominoes, unsurprisingly, the volunteer line to aid the cause was endless and, true to their organization's mission and profile, the Rubber Division members were, to use a military term, "front and center."

In December 1938, David Goodrich, chairman of the board of B. F. Goodrich, faced an unenviable task. During a time when his company and his country had to weigh every decision made as if it might be their last, Goodrich had just dismissed Samuel B. Robertson, his president, and had to look for a new leader. The following summer, Goodrich found an individual who had a solid background in international business, especially in Japan and Germany, and so he sailed on the *Queen Mary* to London to negotiate with Dunlop's John Collyer.

On July 11, 1939, Collyer and Goodrich agreed on terms, primarily dealing with control of the company rather than with money.[5] Collyer had only been in the United States for a couple of weeks when he had to go to Washington, D.C., for a meeting and asked B. F. Goodrich's top research chemist, Waldo Semon, to accompany him. "We finished the meeting," said Collyer, "and Semon asked me if I'd like to meet Senator Truman. I said, 'Who? I've never heard of him.' And he said, 'I think I can get you an appointment today.'"[6] Harry S. Truman, the Democrat from Missouri who of course would become president in 1945, sent word via his staff that he would meet Collyer and Semon in his office that evening at six o'clock sharp for only thirty minutes.

"We were in his office promptly at six, and we told [Truman] what might happen if war broke out. I know the Americans feel like nothing could happen to us, because with an American fleet in one ocean, and the British in another, that would take care of everything." At 6:30, Collyer took out his watch and said, "Mister Senator, thank you very much. And he said, 'What the hell are you talkin' about?' I said the word I had was we could only stay for a half-hour. Truman said, 'I'll change that.' He went into the other room, then came back and sat down. The meeting with our great scientist Waldo Semon and Senator Truman went on until 11:30 that night. [Truman] was on our team." The following week, Collyer was summoned to Washington to tell a Senate committee about his belief in rubber's importance to the nation, especially if the country would become engaged in war. "It left a deep impression with them," Collyer said. "I could see they were ready for the right type of action."

Collyer, a 1917 Cornell University graduate, struck high notes in Washington but also was well accepted by his company's management and staff. By 1940, he had guided B. F. Goodrich to industry prominence in technology and innovation with the introduction of Ameripol and its use in the first tire made with synthetic rubber developed through American ingenuity and produced in the United States.[7] He was proud of Ameripol and the Liberty tire that sold at a 30 percent premium over the normal price for tires.

Whether it was coincidence or not, less than three weeks after Collyer and Semon introduced Ameripol, on June 28, 1940, President Franklin D. Roosevelt declared rubber "a strategic and critical" material to the nation and announced that the Rubber Reserve Company had been formed with seven other corporations to lead in its conservation.

The other corporations formed were the Metals Reserve Company, Defense Plant Corporation, Defense Supplies Corporation, War Damage Corporation, U.S. Commercial Company, Rubber Development Corporation,[8] and Petroleum Reserve Corporation. The RRC provided guidelines for (1) conserving what relatively small amounts of rubber the country already was producing, (2) establishing measures to increase its production, and, most important, (3) creating a synthetic rubber industry. The RRC said it had one million tons of natural rubber in stock but didn't know how long it would last if the nation became involved in the war. Roosevelt also sent a letter to all state governors suggesting a speed limit of forty miles per hour, ostensibly to conserve gas but in reality to conserve the rubber on tires.[9]

The Rubber Reserve Company with its priority list came under the large umbrella of the Reconstruction Finance Corporation (RFC), which was responsible for developing a list of resources—materials and production facilities—needed to equip and maintain an army and air corps of more than one million and support a buildup of troops anticipated to be six million. To that end, $6 billion was allocated to the War Department, almost as much as the nation had spent in that area between 1922 and 1940. That figure, however, would grow by 1941 to nearly $34 billion for the eventual "Victory Program," and to $150 billion after Pearl Harbor. (The final amount would exceed $288 billion, excluding lend-lease funding to other nations).[10]

With a mandate now in place, a network of agencies, activities, and controls began developing to manage war production operations. The concept of a civilian-controlled group was now not only possible, but highly likely. To begin the process, Roosevelt appointed the flamboyant multimillionaire Texas banker Jesse Holman Jones as the first chairman of the RRC.

The Reconstruction Finance Corporation was originally established under President Herbert Hoover's administration in 1932. Under Roosevelt in 1933, the RFC was directed by none other than Jesse Jones.

Jones was a busy man in Washington. In its November 30, 1940, issue, the *Saturday Evening Post* noted, "Next to the President, no man in the government and probably in the United States wields greater powers." Roosevelt made a similar analysis of his RFC director: "Jones is the only man in Washington who can say yes and no intelligently twenty-four hours a day." Jones did little to negate that public image. The son of a Tennessee tobacco farmer, he moved to Texas at the age of nineteen and, through some money from his uncle's estate, opened a lumber business in Houston. In 1940, Roosevelt selected Jones to join his cabinet as secretary of commerce.

Like most men of power, Jones also had a prominent adversary. Eugene Meyer Jr. also held both a large financial portfolio and ego. Meyer graduated Phi Beta Kappa from Yale University in 1895 — only an asterisk in what would become an impressive, financially lucrative, and politically influential career. When he left Yale, Meyer traveled to Europe and immersed himself in the financial industry, working at banks in England, France, and Germany and returned to New York, where he worked in his father's firm, Lazard Frères and Company. In 1901, he left Lazard Frères and through some of his own prudent investments began his own investment-banking firm.[11] By 1915, his total worth was in excess of $100 million. He became interested in mining, railroads, steel, automobiles, and chemicals. In 1921, he negotiated a merger of Aniline Chemical Company and four other chemical companies to form Allied Chemical and Dye Corporation.

One important catalyst to the "relationship" Meyer had with Jones was the fact that Meyer was the first director of the RFC but resigned in 1932 to purchase the *Washington Post*, one of the stallions of an industry he had not yet tamed. On April 9, 1942, Meyer and Jones met by accident at the annual meeting of the Alfalfa Club in Washington, D.C.'s Hotel Willard. Years of verbal one-upmanship came to an ugly conclusion over a hot topic of discussion — synthetic rubber.

That day, Meyer's newspaper had written an editorial heavily criticizing the way Jones was running the RFC. The *Post* editorial read in part, "Mr. Jones fell down rather badly on the job of acquiring and producing sufficient rubber to meet an emergency that we should have foreseen. . . . The chief reason for his failure is a boundless ambition for power that has led to his taking on more jobs than he can successfully manage." Jones read the editorial and became irate. He even fired off a letter demanding that the *Post* retract the statement. He was still upset when he went to the Alfalfa Club that evening and, almost as soon as he arrived in the ballroom, encountered Meyer. After a verbal volley, Jones, age sixty-eight, grabbed Meyer, age sixty-six, and both fell to the floor and scuffled in what must have looked like a lame baseball brawl.[12]

The confrontation was indicative of the tension and emotions of the war as well as the critical nature of synthetic rubber and the importance of the Synthetic Rubber Program. Neither man really did much to damage the other physically or reputation-wise. However, in 1945, after the war ended, Jones resigned from the RFC and returned home to become publisher of the *Houston Chronicle*.

———

On December 19, 1941, less than two weeks after Japan launched its attack on Pearl Harbor, the Rubber Reserve Company called for an initial annual production of forty thousand tons of general-purpose synthetic rubber to be manufactured by five companies: Goodyear, B. F. Goodrich, Firestone, U.S. Rubber (Uniroyal), and Jersey Standard (Standard Oil of New Jersey).[13] The companies also had to sign a patent-and-information-sharing agreement.

Jersey Standard actually played a very important role in the development of the Synthetic Rubber Program. In 1930, Standard had formed a joint venture with I. G. Farben called the Joint American Study Company. It was the design of JASCO to share expertise in the testing and development of new processes in the oil and chemical fields. It proved fortuitous on Standard's part, as it became the holding vehicle for Farben's U.S. patents, which later would include Buna rubber. Standard's major role, in addition to the one as a manufacturer, was as the guardian of the patents and of the subsequent knowledge of Buna. As one of the elite companies selected by the RRC, it would share knowledge that was critical in the development of GR-S.

In their work with I. G. Farben, two Jersey Standard scientists, William J. Sparks and Robert M. Thomas, discovered butyl rubber. After Thiokol and Neoprene, butyl was the third major U.S. contribution to synthetic rubber. Butyl, considered a specialty rubber, has a wide spectrum of applications that overlap natural rubber to a considerable extent. Its unique property is its low gas permeability, which made it ideal for use in inner tubes.

"One of the best things that the Rubber Reserve Company did in those early days was to get the industry people together to decide what type of rubber was to be produced with the patent situation to be handled by a cross-licensing agreement," said Jim D'Ianni, former chairman of the Rubber Division of the American Chemical Society. "Standard Oil of New Jersey owned the basic patents on Buna S, which they contributed to the effort, and all the other companies agreed to cross license any of the developments they had or would be developing as the program went on. This eliminated immediately many problems that would have arisen throughout the emergency, and the Rubber Reserve Company made a very wise policy decision in arranging this agreement which became known as the 'December Nineteenth Agreement' in later years."[14]

B. F. Goodrich's Collyer recalled in subsequent meetings in Washington, "I was first very active in making plans for B. F. Goodrich, but the government . . . wanted B. F. Goodrich to take the entire job. I said that isn't what made America great. . . . It's competition. We should use the four largest companies to do the job. . . . We were on the move." Whether by inference, suggestion, or cajoling, the RRC indeed made its move, even though it knew that the rubber industry had no commercial process to produce synthetic rubber.

In the meantime, the United States wasn't totally void of synthetic rubber. In addition to Goodrich's efforts, which began with a one-hundred-pound-per-day pilot plant to produce Ameripol, Goodyear's Ray P. Dinsmore, Rubber Division chairman in 1927, had developed a synthetic rubber with properties similar to those of Buna N called Chemigum. Goodyear had experimented with the rubber since Lorin Sebrell came back from Germany in 1938 but provided it only to potential customers on a trial basis. However, in November 1940, Goodyear started production of Chemigum in Akron, Ohio, with an initial run of one ton a day. D'Ianni of Goodyear also conducted extensive research on synthesizing a variety of monomers that could be polymerized with butadiene.

D'Ianni's interest and involvement in synthetic rubber was no coincidence. Prior to undertaking his doctoral work at the University of Wisconsin in 1934, D'Ianni had worked for a summer at the Philly Labs of B. F. Goodrich on Bartges Street in Akron. "My work consisted of analyzing samples of reclaimed rubber," he said. "Before being accepted for this challenging task, I was quizzed at some length by Waldo Semon on my knowledge of organic chemistry, including Kekule's theory of the structure of the benzene molecule!"

At Firestone, John N. Street directed that company's efforts in polymerizing butadiene and styrene, and Firestone also built a synthetic rubber pilot plant in Akron for tire applications, but before he did that, he needed some Buna S for testing purposes. The only place to get Buna S was in Germany, so Street asked his associate, Ernest T. Handley, for some help. With a little chicanery, Handley delivered.

With designs of acquiring some Buna S for Street, Handley devised a plan in which he would work off of a relationship with a Swiss supplier to get the synthetic rubber sample. He recalled that a German chemical company was anxious for Firestone's business in Switzerland, where the company had a joint-venture plant in Pratteln. That company was I. G. Farben. Handley told Rubber Division historian Benjamin Kastein:

> And I helped this salesman to get Swiss francs at a time when you could not get any money out of Germany under Hitler. I worked on this project for perhaps six months knowing that Dr. Street needed more than a small sample. The way it worked out, I was invited to Leverkusen and I. G. Farben's research lab where the production of Buna S was carried out. I was able to get a look at how both the Buna N and the Buna S was made. [He always had in mind his objective to return to Akron with Buna S.]
>
> I took along an order for one hundred kilos of Buna N for our [joint venture company] in Switzerland. I persuaded this sales representative to help me get Buna S by taking this order and processing it for Buna N. Then we went out at night, he and I, and we emptied the Buna N out of the drum and filled it with Buna S and left the same marking, Buna N. He saw to it that these particular drums were shipped to Switzerland to our parent company. I did that because I didn't want the name Firestone particularly involved in the transaction. From [Switzerland] we shipped these drums of Buna S directly to Dr. Street.[15]

With the Buna S in his hands, Street had more than enough of the German synthetic rubber to conduct tests on compatibility and wearability and enable Firestone to develop its own synthetic rubber formula. "After

Street did his work," Handley said, "we knew then, that we had a [synthetic] rubber that was compatible."

By the end of 1941, it had become hauntingly apparent that even the expedited efforts of the Synthetic Rubber Program were too slow. Even the early critics of the program now realized that it wasn't a matter of *if* they should support it, but *how* and *how fast*.

If 1941 was the year the Japanese woke the sleeping giant, then 1942 was the year that giant not only ran over the hurdles but also knocked them over in its path—and there were many hurdles, mostly political. While progress had been made since the start of the war, the original fab four plus one of the "December Nineteenth Agreement"—Goodyear, B. F. Goodrich, Firestone, U.S. Rubber, and Jersey Standard—weren't producing synthetic rubber at a pace to keep up with the snowballing demand. Jesse Jones, the RFC's oft-criticized leader, authorized an expansion of the program to 400,000 tons annually in mid-January, then to 600,000 in March. By April the figure would rise to 800,000 tons, including 705,000 tons of Buna S-type rubber.[16]

Aside from the usual production start-up issues with equipment, raw materials supply, and so on, the originally defined copolymer combination of butadiene and styrene as defined by the RRC's companies wasn't the problem that was slowing things down; rather, it was the polymerization. The proper mix or "recipe" had not yet been determined. This was a key element in cloning Buna S, the established German standard. The correct mix also had to be compatible with existing machinery.

On March 26, 1942, the groups, via independent but collaborative research through several information-exchange forums such as those offered by the Rubber Division, developed a "mutual recipe" to produce consistency and a common yield of GR-S. The butadiene-styrene (75 percent to 25 percent) mixture remained intact, but with the addition of a catalyst or initiator (potassium persulfate), an emulsifier (soap), and a modifier (dodecyl mercaptan). Goodyear's D'Ianni said of the mixture, "It turned out to be quite a simple recipe." And he quickly pointed out that the industry also had some help from academia. "They [academic researchers] were the ones who studied the basic mechanisms of polymerization and they developed the analytical methods that were needed in the plants in order to follow the reactions. They were the ones who contributed greatly to determining polymer structure, molecular weight and its distribution and all of the other basic facts that we now know about polymers."[17] With a "mutual recipe" now in place, companies were a little more confident about achieving the targets established by the RRC, although the downside was a pause on company innovations for companies' sakes.

A Goodrich engineer, William I. Burt, who managed the company's Ameripol plant, was selected to chair a committee to design the first four synthetic rubber plants, and colleague Walter Piggott headed the engineering endeavor. Within nine months, each of the plants had begun production. The first "bale" of GR-S was produced by Firestone in Akron on April 26, 1942. On May 18, Goodyear followed with its Akron plant, U.S. Rubber produced its first bale on September 4 in Naugatuck, Connecticut, and B. F. Goodrich was the last to produce its first bale at Borger, Texas, on November 27.

By the summer of 1942, the new synthetic rubber plants were progressing ahead of schedule, and by the end of the year, the four plants had produced more than 2,200 tons of GR-S rubber. By 1945, with the additional support of a network of facilities, the United States would produce 920,000 tons of synthetic rubber, 85 percent of which was GR-S. The first four plants were producing 547,000 tons, or about 70 percent of the nation's output of synthetic rubber.[18]

In spite of the optimistic early start, the administrative issues continued to obstruct what otherwise was a tremendous contribution by the industry and its many participants. Among many obstacles, two of them dealt with what was the best source of butadiene, petroleum or alcohol (i.e., oil industry versus the agriculture industry, and all its related political issues),[19] and who was running the ship—the Reconstruction Finance Corporation and Jesse Jones (and his ego) or the War Production Board, Donald Nelson (and his ego), and his rubber coordinator Arthur Newell?

Tired of all the internal disparagement, Roosevelt called on an old friend to help him decide what was best for the country. On August 3, 1942, FDR selected Bernard Baruch, a financial advisor to several presidents, to chair a three-man committee to resolve the issues. Roosevelt included Karl T. Compton, president of MIT, and James B. Conant, president of Harvard. Compton and Conant were keenly aware of the importance of rubber. Their universities had contributed many great minds to the industry. In addition, in the fall of 1939 in Boston, both men were keynote speakers, along with Goodyear's P. W. Litchfield, at the Rubber Division of the ACS's centenary celebration of Charles Goodyear's discovery of vulcanization.

Unlike his fellow academic committee members, Bernard Mannes Baruch was a southern boy, but as brilliant as any "Yankee" north of the Mason-Dixon line. Born in 1870 in Camden, South Carolina, he and his parents left the South for New York when Bernard was ten years old. However, even into his seventies, he still retained a hint of his southern accent. The son of a German immigrant who had come to the United States in 1855 served as chief surgeon and on Robert E. Lee's staff during the Civil War. Baruch graduated from City College of New York in 1889 and earned three dollars a week in his first job as an office boy running errands in the banking and financial district of Manhattan. However, history would define Baruch not as a wealthy Wall Street financier but as a financial advisor to six U.S. presidents during World War I and World War II.

President Roosevelt knew Baruch and was familiar with his portfolio of service to the nation's chief executives. Roosevelt had even asked for his advice in 1940, prior to establishment of the Rubber Reserve Company. The two were friends and, years later, toward the end of his life, FDR would travel to Baruch's seventeen-thousand-acre South Carolina plantation for some relaxation. This time, FDR's objective and mission was much more focused. He wanted Baruch and his committee to put some order in this Synthetic Rubber Program.

Roosevelt's appointment of Baruch, Conant, and Compton to the Rubber Survey Committee allayed the fears of many, if only temporarily. The media liked the decision too. In an editorial on August 19, 1942, the *New York Times* stated, "Unquestionably the very best formula for synthetic rubber is the one President Roosevelt has hit upon. It comprises Baruch, Compton, Conant. The finest ingredient for producing any mixture is brains."

Within slightly more than a month, on September 10, the Rubber Survey Committee issued its famous "Baruch Report." Before itemizing its recommendations, the report offered an almost macabre analysis of the state of the rubber industry:

> Of all the critical and strategic materials . . . rubber is the one which presents the greatest threat to the safety of our nation and the success of the allied cause. Production of steel, copper, aluminum alloys or aviation gasoline may be inadequate to prosecute the war as rapidly and effectively as we would wish, but at the worst we are still assured sufficient supplies of these items to operate our armed forces on a very powerful scale.
>
> If we fail to secure quickly a large new rubber supply, our war effort and domestic economy both will collapse.

The report, in essence, recommended the introduction of a maximum speed limit of thirty-five miles an hour and a five-thousand-mile annual maximum per car; the expansion of rubber reclaiming efforts; maintenance of an inventory of 112,000 tons of natural rubber; investigations into the use of guayule and cryptostegia (alternative sources of natural rubber); an increase in the annual production of synthetic rubber from 705,000 to 845,000 tons, primarily for tires; the construction and operation of fifty-one plants to produce the monomers and polymers needed to manufacture GR-S; and the appointment of a director with unlimited powers that only the president could override.

The committee recommended William M. Jeffers, president of Union-Pacific Railroad, as the first rubber director, or "czar." The deputy director was Bradley Dewey, president of Dewey and Almy, a chemical company in Boston. Dewey was an ally of the Rubber Division who would become president of the American Chemical Society in 1946. The third man to complete the triumvirate was Assistant Deputy Director Lucius D. Tompkins, vice president of U.S. Rubber. Tompkins was head of the tire division but previously had worked with General Tire, for the most part in their plantation operations.[20]

One of the individuals who was asked to provide input to the new Synthetic Rubber Program leaders was Goodyear's Ray P. Dinsmore, chairman of the Division of Rubber Chemistry in 1927. "I was called to the office of the rubber director to set up a research program which involved industry, and also required the testing and examination of scores of synthetics which were suggested from time to time," he told Rubber Division historian Herbert A. Endres in 1965.[21] "This was an interesting situation because the Rubber Director's office itself was not organized. There was no outline of what my division of the R&D was supposed to do and no personnel to do it. Moreover, we were hustled into an empty building with practically no furniture. I had a desk and two chairs. I was able to obtain some very high caliber men because of the patriotic nature of the appeal, highlighted by the recent Baruch Report on the rubber situation."

Among the individuals Dinsmore mentioned was Robert R. Williams, chemical director of the Bell Telephone Labs; Carl F. Prutton, head of the Department of Chemical Engineering at Case; Joseph Elgin, head of the Department of Chemical Engineering at Princeton; William B. Weigand, research director at Columbian Carbon (who served as Rubber Division chairman in 1923); and Norman A. Shepard, director of technical services at American Cyanamid. Dinsmore said:

> These men, accustomed to first-class equipment and surroundings, and to directing large groups of people, were suddenly dumped into a building in Washington, where they did not even have a chair to sit in. Whenever I returned to my office after a short absence, I was sure to find them roosting there like a flock of chickens. This made me very nervous for fear they might quit in disgust, and I worked hard to get facilities for them.
>
> Meanwhile, we were composing a program and trying to learn about government red tape. Yet this rather temperamental group worked together splendidly. We gradually ironed out our difficulties and I acquired a group of firm friends whom I value greatly. In fact, when I decided I should return to Akron to work on problems with synthetic rubber, to my great surprise, I almost had a rebellion on my hands.

While the efforts of industry were certainly critical to the success of the project, there was still a great amount of knowledge and research that the effort had not yet tapped. In the fall of 1942, the Rubber Research Program was created to extend the study of synthetic rubber to the minds of chemists at several colleges and universities throughout the country.

Bradley Dewey, the newly appointed deputy rubber director, selected Robert R. Williams of Bell Telephone Laboratories to guide the pursuit. In early December, Williams visited eleven universities.

The key individuals and institutions Williams targeted included Izaak "Piet" Kolthoff at the University of Minnesota, Carl "Speed" Marvel at the University of Illinois, William Draper Harkins and Morris Kharasch at the University of Chicago, William Reynolds at the University of Cincinnati, Edward Weidlein of the Mellon Institute, Charles Price at the University of Notre Dame, and many others from Case Institute of Technology (Case Western Reserve University), Princeton University, Cornell University, and, in 1943, the University of Akron.

When rubber is mentioned by those knowledgeable about the subject, there is no individual who transcends the entire timeline of rubber's development like George Stafford Whitby. From the jungles and plantations to synthetic rubber and the products that were precursors to those used in today's modern tire factories to the teaching of courses in rubber, Whitby was the prince of this polymer.

A wiry man with his signature well-trimmed mustache and horn-rimmed glasses, Whitby was born in England in 1887, received his undergraduate, graduate, and doctoral degrees at McGill University in Montreal, Canada, and continued his rise to prominence while calling the University of Akron and Akron, Ohio, his home. Whitby was instrumental in developing measures to ensure the quality of rubber—both natural and synthetic—for most of the eighty-five years he spent on earth. He ignored the advice of his mentor, Sir William Tilden, that it was commercially impossible for synthetic rubber to become a reality, and during World War II, Whitby became a prominent figure in that endeavor, serving Canada, Great Britain, and the United States in the synthetic rubber development field.

His contributions to the U.S. Synthetic Rubber Program administered through the Office of Rubber Reserve, as director of the Chemical Laboratories in the Department of Scientific and Industrial Research of Great Britain, and as chairman of the Canadian Synthetic Rubber Technical Advisory Committee, among many other achievements, are well documented. He authored or coauthored more than ninety-five publications, eighty articles, and fifteen patents on rubber and rubber technology. In layman's terms, Whitby knew rubber inside and out, like no one in his era.

Whitby's first exposure to the Rubber Division came at a relatively early point in his career. "It was at the division's spring meeting in 1921 in Rochester, New York, when I first presented a paper on ultra accelerators," Whitby told division historian Herbert A. Endres in a taped interview on August 11, 1965, at Whitby's Bath, Ohio, home. Whitby said that the paper was coauthored by Arnold H. Smith, secretary and treasurer of the division who would later become its chairman and subsequently a president of Monsanto Canada, Ltd. Smith also recalled in a later interview with Endres, "I think Dr. Whitby was really over[ly] generous in making me a coauthor. He was still conducting research work at McGill and wrote to me that he didn't have the proper equipment to conduct some of the tests and asked me to help him. I told him I would, then he wrote this paper and he really had the ideas. . . . I just carried out his instructions."

However, Whitby continued to take advantage of his association with the Rubber Division and became a regular contributor of technical papers to *Rubber Chemistry and Technology*, the rubber industry's major vehicle of communication. His published works in *RC&T* included "A New Tetramethylbutadiene" and several papers on methyl rubber and the polymerization of isoprene and dimethylbutadiene.[22] He continued to submit papers on a routine basis for many years, sharing his knowledge with fellow members.

Whitby's additional work on synthetic rubber gave him the opportunity to present some of his research efforts to the fall meeting of the Rubber Division in New York City at Columbia University. "I started some synthetic rubber research at McGill University not long after the close of World War I," he said. "The research primarily focused on methyl rubber." The Germans had failed in their attempts to transform methyl rubber into even an adequate substitute for natural rubber, so Whitby decided to see if he could accomplish in his lab what the Germans couldn't do in several war years.

While synthetic rubber always occupied a space in his creative mind, it wasn't until 1942, when Hezzelton E. "Hez" Simmons, president of the

University of Akron and former professor of rubber chemistry succeeding Charles M. Knight, lured Whitby to the campus. Whitby had come from Great Britain, where he was director of the Chemical Research Laboratory in the Department of Scientific and Industrial Research. Simmons had one objective in bringing in Whitby: establish an internationally renowned rubber research organization. On February 1, 1942, the university's board of directors approved the development of a Rubber Technical Institute, and James Schade was appointed director.[23]

In 1943, the institute received a contract for twenty-nine thousand dollars from the Rubber Reserve Company for research into synthetic rubber, and Whitby assumed control. Knowing the relationship of the University of Akron to the rubber industry and to the Rubber Division, Whitby said, "I cannot claim to have been in it during its first few years . . . but I think the Division of Rubber Chemistry has played a very important part and has been very fruitful in advancing the scientific study of rubber in the United States."

At the end of 1942, when the crumb rubber had somewhat settled on the many twists and developments of the continuing saga of the synthetic rubber drama, an apex was reached in Akron, Ohio, at the Mayflower Hotel on Main Street. On December 29, the newly formed Rubber Research Group gathered at a "Secret and Confidential" meeting held by the War Production Board and the office of the assistant deputy rubber director. The subheading on the notice read, "For Research and Development of Synthetics." The forty-two men attending the post-Christmas meeting called it the Copolymer Research Meetings.

Robert R. Williams, the assistant deputy rubber director, was chairman of the meeting. It is not an exaggeration to say that the other forty-one attendees were virtually every top research chemist from every rubber company in the country, including Williams's almost hand-picked researchers from the top universities. The names included many past and future Rubber Division chairmen and several future Charles Goodyear Medalists. The energy from such a high-powered group could have turned even cold, snowy Akron into a tropical paradise.

Jim D'Ianni of Goodyear recalled the meeting in an address to the Rubber Division and the Washington Rubber Group in July 1979.[24]

"The rubber companies reported on their research work on general purpose synthetic rubber and these reports became known as CR reports," he told the groups. "I presented a report, CR-7, with Dr. [Albert M.] Clifford of Goodyear, which was quite an exhaustive one, since it reported work on about two hundred monomers, which we used to replace styrene, hoping to make a superior synthetic rubber. As it turned out, these monomers could make good rubbers, but not better rubbers. . . . We were fortunate that the right decision was made at the very beginning of the program to concentrate on GR-S as the synthetic rubber for the industry."

That meeting and subsequent meetings typified the combined research efforts of government, industry, and academia during World War II, especially for the Synthetic Rubber Program. During the war, the chemists, scientists, engineers, and academicians accomplished a great deal. They improved polymerization processes, produced modifiers for existing machinery, created new carbon blacks, developed new accelerators, developed improved processing equipment, and in general improved development and production techniques ad infinitum.

Suppliers too were critical to the evolution of synthetic rubber in the United States, and one of the contributions included the development of the oil-furnace process for carbon black by Joseph C. Krejci of the Phillips Petroleum Company. Krejci would receive the Charles Goodyear Medal in 1974 for his pioneer work in the evolution of furnace blacks. Monsanto contributed improved accelerators, and many other companies in the reclaimed rubber industry enabled the country to supplement and extend its dwindling supply of natural rubber.

By 1943, fifteen synthetic rubber plants were in operation, and the focus continued on incremental changes to existing processes. In September, Rubber Director William Jeffers was replaced by Bradley Dewey, his deputy. On finding out about his new position, Dewey telephoned his wife and said, "Well, we're moving to Washington now. She said, 'okay . . . if you have any friends see if they can help us find an apartment.'"[25] On his way to Washington from Boston, Dewey stopped in New York to visit his good friend Bernard Baruch. "I said chief—I always called him chief," said Dewey. "How do I start this job? What do you recommend? He said to 'get tough. Just forget that you have any

The Rubber Division members always saved some time for relaxation. At the banquet
for the 1956 meeting in Cleveland, some of the attendees were Dorothy Osterhof;
Harold Judson Osterhof, director of research at Goodyear (who would become the
1971 Charles Goodyear Medal recipient), Leora Straka, librarian at Goodyear
Research; Lois Brock, librarian at General Tire; and Mel Lerner, editor
of *Rubber Age*. (Photo courtesy of the University of Akron Archives).

Jim D'Ianni is the last individual to
have served as both chairman of the
Rubber Division (1964) and
president of the American
Chemical Society (1980). Among
his many achievements both
outside and inside the division, he
is the recognized founder of the 25-
Year Club. (Photo courtesy of the
University of Akron Archives.)

On October 3, 1958, a celebration was held to recognize the fiftieth anniversary of rubber chemistry at the University of Akron. Notables from academia, the industry, and the media were in attendance. They included (*left to right*) Howard Cramer, Mel E. Lerner, and George Hackim. Cramer was division chair in 1949 and worked for Sharples Chemicals in Philadelphia. Lerner, editor of *Rubber Age* magazine, was one of the catalysts in forming the suppliers' cooperative cocktail committee in 1932, the precursor of the 25-Year Club. Lerner won the division's Distinguished Service Award in 1973. Hackim was a vice president from General Tire. (Photo courtesy of the University of Akron Archives.)

On November 6, 1966, the International Rubber Science Hall of Fame inducted Hermann Staudinger. G. Stafford Whitby, a University of Akron legend in rubber chemistry and 1954 recipient of the Charles Goodyear Medal, holds an unframed portrait of Staudinger. University of Akron onlookers included (*left to right*) Alan N. Gent, Herman Mark, President Norman P. Auburn, Maurice Morton, and Whitby. Three of the four would later win the Charles Goodyear Medal—Gent in 1990, Mark in 1988, and Morton in 1985. Jointly founded by the division and the University of Akron's Institute of Polymer Science in 1958, the hall of fame honors researchers who have made a substantial contribution to the understanding of rubber-like materials or have created an outstanding invention. (Photo courtesy of the *Akron Beacon Journal*.)

The past and present come together when Ben Kastein (*left*) and Shelby Washko talk about Rubber Division history. They should know because Kastein was division historian for fourteen years (1978–92) and Washko, the current historian, took over from him.

Akron native Herbert A. Endres, a research chemist at the Goodyear Tire & Rubber Company who held twenty-one U.S. patents, was the first historian of the Rubber Division. Jim D'Ianni, former chairman of the division, presents a Certificate of Special Appreciation to Endres in 1976. (Photo courtesy of Leland Endres.)

Appropriately named the Rubber Band, this group of musicians were members of the Philadelphia Rubber Group, led by Ralph Graff (*far left in the back, holding tuba*). This is a photo from the 1984 banquet in Philadelphia. The band was founded by Graff and Henry Rensberg of Carlisle Tire at the 1960 Rubber Division meeting in Oreland, Pennsylvania. Graff and Rensberg thought that the organization needed some social activity between morning/afternoon golf and the evening banquet. Most of the musicians were members themselves, but often a "ringer" or family member or two would sit in. Rensberg had a part-time job as a music teacher at Dickinson College. Some of the other regular members of the group included Dick Hendrickson, T. W. Elkin, Don Gorman, Al Larsen, A. E. Lawrence, Tom Loser, and Charlie Glaser.

The Steering Committee continues to be a key driving force behind Rubber Division operations. This contingent met in Cleveland in October 1975. *Front row:* H. W. Day and Fran M. O'Connor; *back row:* E. M. Dannenberg, A. M. Gessler, W. Curt Rowe, Ralph Graff, Frank Floren, and Dan Meyer. (Photo courtesy of Ralph Graff.)

The tradition of using a bagpiper to lead the entrance of the officers into the banquet started about 1955 at the Philadelphia Benjamin Franklin Hotel. The original purpose was to get those at the head table—the officers, area directors, Charles Goodyear Medalist, and VIPs—from their cocktail reception to the banquet. The tradition ended in the mid-1980s but was resurrected at the 2009 spring banquet in Akron for the centennial celebration. This photo is from the 1984 banquet. To the right and immediately in front of the bagpiper is Ronald J. Ohm, ASARCO, secretary of the division 1984–85.

Marge Bauer (*left*) works with Connie Morrison-Koons in one of their many tasks at the Rubber Division's offices on the University of Akron campus. Bauer was the first full-time "employee" of the division who originally was assigned to handle the office of the treasurer of the Rubber Division of the ACS from the offices of Harwick Chemical. Morrison-Koons was Bauer's first assistant. (Photo courtesy of Marge Bauer.)

Ernie Zielasko, founder in 1971 of *Rubber & Plastics News*, received the Rubber Division's Distinguished Service Award in 1985. Zielasko was a longtime supporter of the division and its contributions to the rubber industry. (Photo courtesy of David Zielasko.)

In 1976, the Rubber Division dedicated its library to the memory of John H. Gifford, division chairman in 1971. Gifford, who was employed by Continental Carbon in Houston, Texas, died in a tragic automobile accident in 1975. (Photo courtesy of University of Akron Science and Technology library collection.)

manners. We told the President and Congress that you should have absolute power and should not be subordinate to anyone. Clean up and fire people that are on the fringes of the Synthetic Rubber Program.' I said, 'what do you mean?' He said, 'Be a son-of-a-bitch!'"

Relatively late in the effort, the Synthetic Rubber Program and its many administrators realized that quality control, particularly at the plant level, was a growing concern. From a timing standpoint, labs couldn't be established until the plants were built, and those were continuous works in progress. Two labs, or laboratory efforts, already existed—Bell Telephone Laboratories and the National Bureau of Standards.

On February 2, 1944, before he left his position, Dewey added another lab in Akron to work more closely with the new plants to serve as an objective mediator of testing of GR-S and to evaluate new polymers. The lab was administered under contract to the U.S. government by the University of Akron but operated separately from Whitby's university laboratory. The laboratory and pilot plant were located on West Wilbeth Road, near the government-owned copolymer plant operated by Firestone. The two-story structure had eight rooms on the first floor (for milling and curing) and eleven rooms on the second (for chemical testing, physical testing, abrading, and aging rooms). The lab was officially dedicated on June 28, 1944, and Dewey; Hez Simmons, president of the University of Akron; E. R. Gilliland, assistant to Dewey; and Stanley T. Crossland, vice president of the Rubber Reserve Company were in attendance. In his dedication address, Dewey said that the nation's scientists are committed to producing a better synthetic rubber "than any tree rubber ever produced" and everyone involved in the program "will have a far better understanding of synthetic rubber."[26] James A. Schade, a charter member of the Rubber Chemistry Section in 1909 and formerly director of research of the B. F. Goodrich Company, was the first manager of both the pilot plant and laboratory. He reported to Simmons. The plant was sold to its nearby neighbor, Firestone, in 1956.

Dewey became tired of the out-of-character persona that he had to assume as rubber director, left on July 25, 1944, and returned to his firm, Dewey and Almy. Before he left the office, the first butadiene-from-alcohol plant began production in January 1943. While he even recommended abolishment of the position of rubber director, Dewey's efforts were typical of the other true heroes of World War II, but many of those unsung heroes took the form of lab technicians, compounders, researchers, administrators, and support personnel. It took a concerted effort on the parts of millions of people.

The war had an impact on the Rubber Division meetings as well. The 1944 Rubber Division chairman, Harry Gray of B. F. Goodrich, recalled in a 1965 interview with division historian Endres, "I arranged to have the meeting called off. Our trains were needed for the war effort rather than having a bunch of rubber technologists travel to New York for a meeting." In 1945, all the general meetings were postponed. During this period, it was common for issues to be discussed on the telephone and for decisions to be made via mail balloting.[27]

Willis A. Gibbons, who was the first chairman of the New York Rubber Group in 1928, has been the only Rubber Division chairman to serve two terms since 1919, and he did that in consecutive years in 1945 and 1946. He said there was a logical reason for that and the dearth of meetings during the war. "Every division of the American Chemical Society (ACS) had to 'suffer' under a two-year term of its chairman because of the war," Gibbons told Endres. "The government had forbidden meetings of societies of more than one hundred persons. When the order came out, we were notified by the ACS that each chairman would continue in office until the next year." Gibbons praised the efforts of the division during the war. "Although we missed a lot of meetings, there was still a great cohesiveness between every one. We missed seeing each other, but we knew we had to help the war effort," he said. "During this time, we had a most extraordinary executive committee who traveled long distances to have meetings in lieu of the big meetings."

The work was nonstop for the members of the Rubber Division. Long hours were not expected, they were just routine. The only thing that members didn't have was a staff or general office area to accumulate material for the common good, but that didn't prevent them from contributing. Their division offices were their homes. Their spouses were their support groups in more ways than one. The Synthetic Rubber Program had spawned a great deal of information sharing and collaborative activity, and many individuals worked tirelessly for little recognition and little glory—rewards the free enterprise system routinely offers high-achieving individuals today. The few meetings held during World War II were welcomed respites from reality.

As in all things in life, there are critics. Those who have a different opinion about the Synthetic Rubber Program often say it was a "mess." No one will disagree that for a period of time the effort indeed appeared to perform like a car with bald tires on an ice rink. However, numbers and truly monumental efforts belie the naysayers. In 1942, the United States produced 3,210 tons of GR-S. The following years' figures reveal 182,250 tons in 1943, 670,268 tons in 1944, and 756,042 tons in 1945, the last full year of production.[28]

The qualitative results are more impressive but inherently more difficult to count. The Synthetic Rubber Program didn't just produce a tangible product, it produced creativity, innovation, and a work ethic that is still the envy of several generations. The successes that were due to the heroic efforts of members of the Rubber Division continue to manifest themselves even today. The division was a microcosm of a great partnership among the government, the rubber industry, and its related industries, colleges, and universities.

In the October 1966 issue of *Rubber World*, four men, Otto Scott, Bill Mulligan, Joseph Del Gatto, and Ken Allison, reflected on the state of the nation and the challenge faced by the rubber industry during World War II: "The creation of an industry is far greater and more complicated a task than even the most highly dramatic instances of individual discovery. It is unbelievably complex, involves so many men and factors, is so intricate and intertwined within the national context, that it is virtually untraceable."

———————

In 1945, first on May 7 and then on August 15, the war that began for Americans on December 7, 1941, at Pearl Harbor had come to an end. While there was celebrating in the streets of large cities and on farms, fields, and every other conceivable form of geographical structure around the world, the joy would soon be tempered with the reality of a great rebuilding process. With the exception of Pearl Harbor, the United States had escaped devastation to its own land, but at the cost of thousands of lives.

Some critics might question if the Synthetic Rubber Program would have survived without the Rubber Division. Probably. However, the important question is, would the Synthetic Rubber Program have survived without key technical knowledge? Probably not. Toward this objective, the groundswell of information that the Rubber Division offered its members in advance of the war certainly helped them achieve the most desirable results. In the meantime, the rubber industry and the members of the Rubber Division returned to their prewar activities, but with a different focus. Now they were intent on using their knowledge of synthetic rubber toward a different objective—one that focused on rebuilding private enterprises.

PART FIVE

ADOLESCENCE

A
RENAISSANCE

The world will never have lasting peace so long as men reserve for war the finest human qualities. Peace, no less than war, requires idealism and self-sacrifice and a righteous and dynamic faith.

JOHN FOSTER DULLES,
U.S. SECRETARY OF STATE, 1953–59

The end of World War II precipitated an expected but unprecedented global euphoria. It also gladly compelled former participants, observers, and even the bystanders to divert their attentions in other directions. By the end of the war in 1945, the United States was spending almost sixty-five billion dollars on defense — approximately 37 percent of the gross domestic product and nearly 90 percent of all federal spending.[1] Economic historian Alan Milward wrote, "By 1945, the foundations of the United States' economic domination [for] the next quarter of a century had been secured. . . . [This] may have been the most influential consequence of the Second World War for the postwar world."[2] During World War II, the efforts of many were focused on just getting things done, period. The government's Synthetic Rubber Program is one major example. However, postwar issues revolved around funneling energies and capital into nonpugilistic endeavors and reclaiming private enterprise.

Americans certainly anticipated focusing on things other than war, but the reality was that they first had to deal with the remnants of World War II—the obligations—before they could even begin to investigate their very large wish lists. Fortunately for the many people and organizations that were responsible for accomplishing the first item of order, the government would return the favor. With the implementation in 1942 of the Rubber Research Program, the government already had taken one giant step forward. The second item would be handled with relative ease because the time period after World War II would provide an open forum for pervasive creativity. With the government maintaining control over material and intellectual property during the war, there was a great pent up demand for good old American ingenuity. The postwar years were an exciting time to be involved in any area that begged for originality and imagination—research, development, production, sales, and almost any profession. The entire world was an attentive market. Americans were war weary and anxious to reacclimatize themselves to more pleasant and less stressful times. Nations were rebuilding, and most were both mentally and physically devastated and looking to the United States for every conceivable form of help. For those with resourceful minds who were ready to handle the demands, Uncle Sam's innovation incubator was warming up.

Because of its major participation in the war effort, the rubber industry was one of those businesses concerned with readjustment issues, particularly as they related to creating materials and products for new, non-war-related markets. Many members of the Rubber Division of the American Chemical Society and the companies that had made significant contributions to the Synthetic Rubber Program established by the Rubber Reserve Company in June 1940 were suddenly aware that their portfolios of talent and contributions, while impressive, needed to be readjusted. Natural rubber, the staple of the industry before World War II, was relatively dormant for several years, and the synthetic rubber industry that participants had helped to create yesterday was today a dichotomy.

The expected resurgence of natural rubber didn't materialize, and pending a decision from Congress on the status of the synthetic rubber plants, this meant that there was no foreseen antidote to the almost overnight burgeoning addiction to synthetic rubber that the program itself had created. The United States was caught between, on the one hand, a loyalty to its allies—primarily Great Britain—who had vested interests in natural rubber and, on the other hand, its responsibility to the U.S. taxpayers for jobs and a commitment to reinvest the surplus of funds wisely. And Americans certainly needed jobs.

According to the U.S. Army Center of Military History, between eight and thirteen million Americans were in uniform at the end of World War II. During that time, while millions of male workers were on the *front* lines, more than six million American women worked the *assembly* lines that produced goods from the smallest of rubber and metal parts to huge aircraft on a "temporary basis."[3] The *Encyclopedia of American Economic History* credited the legendary "Rosie the Riveter" movement with helping push the number of working women at the war's peak to twenty million, a 57 percent jump from 1940 to 1944.[4] By 1950, that figure would drop to fewer than sixteen million.[5] What was the nation going to do about these sudden changes in the work force and how could the rubber industry participate?

Many members of the Rubber Division were also asking these questions about reentry into free enterprise, but one of their many concerns included the intent of Congress to dispose of those fifty-one synthetic rubber plants established for the war effort and representing an investment of almost $700 million.[6] Most of the division's members had played an active role in that historic and sometimes controversial Synthetic Rubber Program, and in the development of the plants that had shaped and had grown the products of a new synthetic rubber industry.[7] For many companies, employees, suppliers, and vendors, their futures— or fates—rested in Washington, D.C.

As with many things in a capitalistic society, when all the contributing factors are evaluated and weighed, the government will often step in and provide a decision—whether right or wrong. In the period of five years from 1948 to 1953, there were three such exploits that helped shape the rubber industry. These actions also helped define the path that synthetic rubber and the Rubber Division of the American Chemical Society would take into the future. They were the Steelman Report, the Rubber Act of 1948,[8] and the Rubber Facilities Disposal Act of 1953, but all were related to disposition of the synthetic rubber plants.

On August 27, 1947, John R. Steelman, a White House aide to President Harry S. Truman, issued the first volume of a report to the President's Scientific Research Board (PSRB) titled "Science and Public Policy: A Program for the Nation," which provided an analysis of the federal and nonfederal research systems.[9] Steelman's objective in his report was to provide a process that would enable the government to manage its own research and development programs and coordinate research activities of the government, industry, universities, and other institutions.[10] This came on the heels of Truman's vetoing a bill on August 6, 1947, that would have created the National Science Foundation. (One other interesting disclosure of the Steelman report revealed that in 1947 there were only 137,000 scientists and research engineers in the United States.)[11]

Steelman's initial report precipitated the Rubber Act of 1948 (Public Law 469, approved May 31, 1948), which stated, "It is declared to be the policy of the Congress that the security interests of the United States can and will be best served by the development within the United States of a free, competitive synthetic rubber industry. In order to strengthen national security through a sound industry, it is essential that government ownership of production facilities, government production of synthetic rubber, regulations requiring mandatory use of synthetic rubber and patent pooling be *terminated* whenever consistent with national security as provided in this act." The act established that firm policy, but the buildup of the Korean War (from about 1949 to 1953), caused expected delays in implementation. As a result, on January 14, 1950, President Truman signed a bill extending the Rubber Act of 1948 to June 30, 1952. However, on August 7, 1953, another bill, the Rubber Facilities Disposal Act (Public Law 205) passed, which helped to complete the sale of the remaining plants by 1955.

The Rubber Facilities Disposal Act of 1953 would eventually give some closure to a lengthy process of returning the fifty-one synthetic rubber plants back to private industry. It looked a lot like the Rubber Act of 1948, with its disclaimer of guarding national security interests; establishing a free and competitive synthetic rubber industry; protecting small business interests; and getting full, fair market values from facilities. The report was written by the original team that developed the plan (a.k.a. the Baruch report) to create the rubber "czar," among other things, at the end of the war. Its authors were Bernard M. Baruch and other rubber industry allies, James. B. Conant, president of Harvard, and Karl T. Compton, president of MIT.[12]

In a retrospective report in 1976 for the U.S. Energy Research and Development Administration, Paul R. Samuelson wrote, "The striking feature of postwar operations of the synthetic rubber industry was the nine-year lag from 1945 until 1954 in implementing the disposal of the majority of the plants."[13] The report says that the Inter-Agency Policy Committee on Rubber executed a series of small plant sales from 1946 to 1949 "without a strong commitment to a particular future of the sectors of the synthetic rubber industry."

In 1919, when William C. Geer, chief chemist at B. F. Goodrich, hired Harry L. Fisher from Columbia University to join his staff, a friend of Fisher's thought it strange that Geer had delved into academia for industry help. Norman A. Shepard, a close friend of Fisher and a professor of chemistry at Yale University, remarked, "I think that started a population explosion of chemists in the city of Akron."[14] That same year, Shepard left his position at Yale for Firestone.

If Akron and the industry experienced a proliferation of chemists during and after World War I, then the era after World War II, the sequel to Shepard's comment, could be classified as the era of researchers. One great difference between the rubber chemists' involvement in the years

before and after World War I versus World War II was that in the first conflict, for many reasons, the chemists, scientists, and researchers were on
the periphery of the confrontation. Shepard admitted during that era the
industry was just making a transition from an empirical to a scientific study
of rubber. (Charles Goodyear's work was a classic example of an empirical study.) However, because of its vital importance to the efforts, the rubber industry was thrown virtually onto the front lines during World War II.
While the financial investments (i.e., gains or losses) involving the Synthetic Rubber Program can be debated continuously, the often-overlooked
point of the almost overnight buildup of the program and supporting network and facilities are the advancements in synthetic rubber itself.

To aid the postwar research efforts, leadership of the Research and
Development Branch within the Office of Rubber Reserve was reor-
ganized in 1946, and it was likely coincidental, but of the four individuals who were selected to serve as research section chief on a four-month
rotation, two were former chairmen of the Rubber Division. They were
Harlan L. Trumbull (division chairman in 1937), representing B. F.
Goodrich; Morris G. Shepard of U.S. Rubber; and John N. Street (division chairman in 1942) of Firestone and Albert M. Clifford of Goodyear.
However, in August 1947, the rotation leadership ended.[15] In the
interim, Jim D'Ianni of Goodyear served with this group in 1946–47 as
chief of the Polymer Research Branch.[16]

D'Ianni, in a 1979 meeting of the Washington Rubber Group, reminisced about the two-year leave of absence he took in 1946 from his job
at Goodyear to work for the government in the Office of Rubber
Reserve. He recalled his visits to universities, which were contracted by
the government to assist in research efforts with his office. "I remember
very well the excellent meetings we had with professors like Speed
[Carl] Marvel, Piet Kolthoff, Sam Maron, Peter Debye, G. S. Whitby
and many others. We would meet to review the results of industrial
research and academic research and thus good communications were
established between industry, government and academic people."

To the critics of the Rubber Reserve Program, D'Ianni referred to his
legendary colleague. "In support of my opinion," he said, "I would like
to quote from Professor Whitby's book, *Synthetic Rubber* . . . on the government Synthetic Rubber Program:"

The Government-sponsored research and development program has been conducted in an enlightened spirit. On the strictly research aspects of the subject, the workers have been allowed almost complete freedom in their choice of topics, always of course with the implicit understanding that the topics have a bearing on the advancement of knowledge on the subject of synthetic rubber. There has been a completely free exchange of information through the distribution of reports, through conferences held at intervals and through personal contacts among workers. The present writer cannot well imagine conditions more congenial and stimulating to the advancement of knowledge in a specified, broad field than the brotherhood of research workers that constitutes itself in circumstances such as those outlined. Even the most independent and free-ranging research mind derives stimulus of positive value from membership.

Regardless of the many chess moves on the administrative side of the equation, the government's commitment to synthetic rubber research was manifest in its appropriation of $3.5 million annually toward those efforts.[17] The next several years would serve as a platinum period for research and research programs. It was a sign that the bottle of creative juice was ready to be poured and a fictional god of synthetic rubber was preparing to say, "Go thou and create!"

The goal, of course, was to develop a better GR-S, a synthetic rubber that could rival natural rubber in both quality and price. While price had certainly tilted in favor of synthetic rubber during World War II (fixed at 23.5 cents per pound vs. seventy-three cents for natural rubber), no one knew where the market would go in the future or when quality would be improved.

In 1915, Carl Shipp Marvel knew for sure when he graduated from college at Illinois Wesleyan that he was going to be a farmer. "I didn't really consider anything else," he said years later in an interview with Rubber Division historian Herbert A. Endres.[18] Marvel was living at the Gamma Alpha house just off campus at the University of Illinois in Champaign. Gamma Alpha Graduate Scientific Society is a nonprofit fraternal organization that fosters interdisciplinary dialogue among graduate students. As a doctoral student in organic chemistry, Marvel had a very heavy schedule and always worked late at night in the lab, often "until one or two in the morning," he admitted. He would try to get as much sleep as possible then sprint to the breakfast table before it closed at 7:30 a.m. "My classmates gave me the nickname 'Speed' because otherwise I was rather deliberate," he laughed.

From that time in his life until he took his last breath in 1988, Marvel was always called "Speed." Not Carl. Not doctor. Not Marvel. Speed was "Mister Chemistry" at the University of Illinois for forty-one years, from 1920 to 1961. He has also been called the "Father of Synthetic Polymer Chemistry." During World War II, Speed's contributions to the synthetic rubber program were many and significant. However, in 1946, he was one of several prominent rubber chemists from academia and industry to travel to Germany to investigate that country's use of Buna S during World War II to see what the United States and its allies might learn. Others on Marvel's team were Harlan L. Trumbull of B. F. Goodrich, John N. Street of Firestone, George M. Vila of U.S. Rubber, and Al Clifford of Goodyear. "The principle thing we learned," he said, "was the redox polymerization technique which had been developed by W. Kern and W. Becker at Leverkusen [Germany] at the direction of Dr. [Erich] Konrad [of I. G. Farben]. It was this redox process that we used as a basis for establishing the cold rubber process in America."

Cold rubber is appropriately named because it is synthetic rubber that is polymerized at 41 degrees (F) instead of at 122 degrees, the "normal" temperature to manufacture GR-S. The ability to achieve polymerization at the lower temperature gives the synthetic rubber enhanced qualities, especially in such applications as tire treads, and yields a more flexible rubber with greater wearing and weather-resistance qualities. The Germans used benzoyl peroxide, but it was very volatile and exploded a few degrees above water's boiling point. U.S. rubber chemists discovered that cumene hydroperoxide provided better, less dangerous results.

Cold rubber was the product of extensive research conducted by the four major tire companies and Phillips Petroleum. In 1948, similar to the World War II "mutual recipe" for GR-S, cold rubber received like treatment with a "common cold rubber recipe" developed by William Reynolds and Charles Fryling of Phillips. In 1949, at the recommendation of the U.S. teams, the Office of Rubber Reserve authorized instal-

lation of refrigeration equipment on each of the reactors in the copolymer plants that eventually increased production to approximately two
hundred thousand tons a year, equivalent to 47 percent of the general-
purpose synthetic rubber compound in 1948.[19] By 1955, when the United
States had completed the sale of all of the synthetic rubber plants, about
75 percent of the capacity was capable of producing cold rubber.[20]

In a speech delivered to the Washington Rubber Group in July 1979
that reminisced about the World War II Synthetic Rubber Program, Jim
D'Ianni said, "The development of cold rubber was one of the major
improvements of GR-S. . . . Several factors contributed to its success, and
the combination of these factors made cold rubber considerably superior to hot GR-S."

In a not-unexpected but certainly unwanted move, in June 1950 Communist forces invaded South Korea. Barely five years removed from the
end of World War II, many nations were again holding their collective
breath and placing worthwhile projects on hold. The disposition of the
synthetic rubber plants was one of the detainees. The war also prompted
the government to extend the 1950 deadline of the 1948 Rubber Act to
1954—a brilliant move.

The new war notwithstanding, cold rubber had a companion in the
new development market—oil-extended rubber and the subsequent
decade-extended lawsuits. Oil-extended rubber, sometimes referred to
as oil master-batched rubber, was a product of the Korean War and the
shortage in rubber that it precipitated. The process of adding oil to
GR-S (on a four-to-one mixture of cold rubber and "selected mineral
oils") in the polymerization stage essentially extended the use of the synthetic rubber by 20 to 25 percent and increased the low-temperature flexibility (particularly important to cold rubber) and resilience. Sounds
simple, but only courts can complicate things.

In November 1950, General Tire's William O'Neil told the Reconstruction Finance Corporation that his company had a process that
could help solve the rubber shortage. General was not a contract partner of the RFC's research program at the time but would offer the
process to the RFC for a fee. As expected, the RFC declined, but by 1951
Goodyear said it too had developed an oil-extended process via its gov-

ernment research program. The only catch was that General had not yet filed for a patent. Over the course of literally more than two decades, the very bitter dispute involving almost all of the major rubber companies was settled. In June 1974, the U.S. Supreme Court refused to hear General's last appeal of November 1973 and the court of appeals decision. In spite of the legal battles, oil-extended rubber performed well, as advertised by all participants. Goodyear's Jim D'Ianni said, "Oil-extended rubber has been a boon to the whole industry and many millions of dollars was saved by the extension of rubber with oil. Today it represents a significant amount of all the rubber consumed."

The progress that was made in the development of cold rubber and oil-extended rubber during the course of the war and the war's own consequential temporary synthetic rubber shortage were noteworthy, but the developments also likely added fuel to the government's inclination to spend almost $23 million to reactivate many of the same plants they were trying to sell.[21] However, that did not minimize any post-World War II efforts involving research-related projects. The significant achievements of the postwar epoch were only a vignette of the full-length research film, because the stars and "extras" were many. There were a proportionate number of entities involved in the era, among them the entire spectrum of the supply chain from concept—research, development, testing, and so on—to the final products. This included not only the basic raw materials, the chemicals, but also the additives, the reinforcements, and the components that elevated GR-S into a highly successful elastomer that is in use today in hundreds, thousands, and maybe millions of applications.

While the war (or at least the ostensible "battles") would end in 1953, the research train would continue to roll from 1953 to 1961, as no less than five rubber or rubber-related companies made some of the most significant contributions to the industry. They were Firestone and Fred W. Stavely's "coral rubber" (cis-1, 4-polyisoprene), B. F. Goodrich and Sam E. Horne's contribution of Ameripol SN (synthetic natural), Phillips Petroleum and Robert Zelinski's analogue of natural rubber (which went commercial in 1959), Shell's Lee Porter with isoprene rubber in 1959, and Goodyear's Budene (in 1961) and four months later Natsyn. Both Stavely (in 1972) and Horne (in 1980) would eventually receive the Charles Goodyear Medal. Horne's contribution to Ameripol SN was the result of his attempt at producing an analogue of natural rubber with a Ziegler catalyst.

Karl Ziegler was a West German researcher who in 1951 developed a new class of polymerization catalysts primarily used for ethylene. Ziegler's work was improved upon by Italian Giulio Natta, and their combined efforts led to a new class of elastomers to be commercially introduced in the United States in the 1960s.[22] The class of rubbers is called ethylene propylene diene monomers (EPDMs), and they are used in a variety of applications such as roofing materials and weather-stripping in both automotive and building industries and in a variety of applications that require ozone-resistant materials.

By 1953, two major contributions to the industry, cold rubber and oil-extended rubber also had kept the prices of synthetic rubber (about twenty-three cents a pound) competitive with natural rubber (about thirty cents a pound)[23] and the production of GR-S reached about 672,000 tons a year. The other "cold" element was now the Cold War, which began with the Soviet Union, and the Communist empire's presence probably helped the United States to procrastinate about settling the issue of returning the synthetic rubber plants to private industry, operation, and ownership.

While it may have been only a minor irritant instead of a thorn, the disposition of the plants was still a concern and no one involved could move forward unless some decisions were made. Between October 1946 and December 1948, eighteen of the fifty-one original synthetic rubber facilities were closed—even before the 1953 Rubber Producing Facilities Disposal Act went into effect.[24] In November 1953, under the guidance of Leslie Rounds, former vice president of the Federal Reserve Bank, the commission created by the Disposal Act had convened and presented a plan to Congress. Their recommendation, wary of other outside pugilistic threats, had a ten-year "escape" clause attached to the sale of the remaining twenty-nine plants that enabled the government to temporarily reclaim a facility if an emergency (like a war) occurred. Advertisements appeared eight days later, and the deadline for bids was set at May 27, 1954. By that time, seventy-five proposals had been received from thirty-five bidders—eleven alone for the Los Angeles styrene plant—but none for the Institute, West Virginia, facility. On December

16, 1954, the first contract was signed, and by December 27, twenty-four of the twenty-nine plants had been sold, raising more than $260 million.[25] In March 1955, Congress approved the sale of the plants, and the Baytown, Texas, plant, originally passed over because of low bids, was sold to Union Carbide after a second round of bids.

In May 1955, as GR-S production hit 997,000 tons, Congress agreed to continue government-funded research until 1956, reinforcing, for at least one more year, their commitment. With a one-year reprieve granted on the research kitty, many efforts, primarily university laboratories and labs initiated by the government, prepared for the inevitable. The laboratories Marvel ran at Illinois and Avery ran at MIT didn't produce any industry-changing activities, and in 1956 the Akron Research Lab on Wilbeth Road was sold to the Firestone Tire and Rubber Company. Firestone's absorption of the Akron Research Lab was reflective of a shift in prominence of research efforts from the government-owned labs and academia to the major companies, which were growing every year. In 1957, ironically, at the height (at that time) of GR-S production (1.5 million tons per year), the government research program ended.[26]

John M. Ball, Rubber Division chairman in 1955, said that nothing silences a crowd in a meeting better than the words "This paper is now open for discussion." But he had an idea on how to remedy that. One of the bright spots of the division's post–World War II and post–Korean War accomplishments included an increase in the quality of papers presented at its meetings. "I recall the fall meeting we had in Philadelphia," Ball told division historian Herbert A. Endres. "The papers presented that year were, I believe, some of the most outstanding in my career. There was a great deal of new information, and, in particular, Fred Stavely's paper was the best paper I have ever heard in my life."

Stavely, the Rubber Division chairman in 1950 who would receive the Charles Goodyear Medal in 1972, presented a paper on "Coral Rubber a cis-1, 4-Polyisoprene."[27] In addition, Frank Schoenfeld of B. F. Goodrich presented three papers. Local Arrangements Committee chairman Ben Garvey orchestrated the event as Ball introduced Ray P. Dinsmore of Goodyear as the 1955 Charles Goodyear Medalist.

Ball also said that the division initiated two customs in that meeting that remain in effect today. The ideas came from two DuPont men— Seward Byam, division chairman in 1953, and Art Neal, division secretary from 1954 to 1957. "Seward and Art gave me a most excellent indoctrination of the customs of the division," said Ball. "Byam gave me the idea of invited papers. The first ones were in Detroit at the spring meeting when Ernest Bridgewater [division chairman in 1932 from DuPont] spoke on profits and Harry L. Fisher [division chairman in 1928] delivered a paper on educational activities of Local Rubber Groups." The other idea came from Neal in 1954. "He said we ought to recognize the best paper at each meeting," Ball said. "So we put together a committee of two people who were our first judges."

The criteria the judges had to evaluate were (1) the subject matter, (2) the oral presentation, and (3) the visual presentation. "Then we would position the man who was judging the presentation in the back," Ball said, "and the one who evaluated the subject matter would be in the front. We intentionally put him up front to start asking the questions," he added with a laugh.

———————

The International Rubber Science Hall of Fame, the October 3, 1958, progeny of the Rubber Division of the American Chemical Society and the University of Akron, established two major criteria for induction consideration: (1) Its candidates (representing an international flavor), must have conducted technical, scientific research in the field of rubber, and (2) the nominees must be deceased. Maurice Morton, one of the founding academicians of the organization, who also founded the Institute of Rubber Research,[28] told Rubber Division historian Benjamin Kastein in 1981, "Some years ago, one of our librarians felt very put out that she didn't see my picture on the [hall of fame] wall. And I pointed out to her that I might meet one of the requirements, but I could hardly meet the other one."

The inspiration for the International Rubber Science Hall of Fame came from University of Akron president Norman P. Auburn in an April 11, 1958, letter to the committee organizing the fiftieth anniversary celebration of the start of the first courses in rubber chemistry.[29] It is said Auburn had a discussion with Ben Maidenburg, editor and publisher of the *Akron Beacon Journal*, who asked how he could help Akron recognize the great worldwide rubber chemists. Auburn then met with Morton and G. Stafford Whitby, who took the planted seed to Ray Dunbrook, the Rubber Division chairman.

The division agreed to cosponsor the endeavor, and in 1958 the members inducted the first hall of fame class, which included Charles Goodyear, Henri Bouasse, Ian I. Ostromislensky, Carl O. Weber, and C. Greville Williams. Some twenty years later, in his interview with Kastein, Morton insisted that Whitby, his former college professor at McGill and predecessor at the University of Akron, gave them the inspiration for the Rubber Science Hall of Fame. "I would say that the idea was basically that of Dr. Whitby's, who was then professor emeritus," Morton said. "He said that it was high time that the university took the step of having a hall of fame — a rubber hall of fame."

In 1909 the India Rubber Section started with twenty-eight pioneers — men who dared to be different. By 1919, this courageous group still numbered fewer than fifty, though it would climb to slightly more than five hundred just before World War II. By April 1946, it had increased its membership to more than twelve hundred but had visions of an increase not only in quantity but also in quality. The veterans of the division, several of whom had been witnesses to two world wars, now looked toward the younger generation of chemists and technicians to get them into the second half of the twentieth century.[30]

"We had already seen an increase in the number of papers given," said Fran O'Connor of Harwick Chemical, who began his involvement with the division in the early 1960s and became division chairman in 1974. Because he would traditionally lead the group in a premeeting prayer, O'Connor was the division's unofficial chaplain and was often referred to as "Father." "But we needed some continuity in leadership." O'Connor was encouraged to join the division when he went to work at Harwick for Bud Behney. "[Behney] told me, 'I think you'd make a good secretary, and since I'm chairman-elect I think you would have support of the division and, of course, your company.'" Behney was president of Harwick.

Behney's office at Harwick was looking more like an office of the Division of Rubber Chemistry—which it was, in reality. As Behney and his fellow division members prepared to enter the tumultuous 1960s, they too were looking for some radical changes.

CAPITAL  INVESTMENT

A library book . . . is not . . . an article of mere consumption but fairly of capital, and often in the case of professional men setting out in life, is their only capital.

THOMAS JEFFERSON

Benjamin Franklin was the eighteenth-century version of Mister Wizard, a popular youth-focused tele-scientist originating in the 1950s.[1] Franklin became prominent for a cornucopia of ideas and inventions, and of course there was that thing with a kite, a skeleton key, and lightning. However, among his many noteworthy achievements was his contribution to the founding of the Library Company of Philadelphia in 1731. Franklin and some of his scholarly friends needed a forum to discuss current events and so they formed the Junto Society in 1727.[2] The society was a private "fraternity" to brainstorm, among other things, publicly beneficial ideas. One of those ideas materialized into the Philadelphia Library Company, which became the first known subscription library and likely the first truly public library where members could borrow books. The library is still in existence today as a nonprofit, independent research entity.

Slightly more than two centuries later, the Rubber Division of the American Chemical Society also had an educational and research-related inspiration similar to Franklin's. For the first time in a few decades, members of the division who had made significant contributions to its success suddenly faced a new challenge: after two world wars in roughly the first half of the twentieth century, how to focus their combined research and development efforts on peacetime endeavors. In order to address the new opportunities, those rubber industry colleagues felt compelled to take advantage of the many years of research they had already invested. Just like Franklin and his friends, they needed a library.

The rubber chemists and researchers required a universal conduit of technical information. Each of the companies in the rubber industry, primarily the major tire companies in Akron, had reserved areas where valuable information was stored; however, the scientists had grown accustomed to *sharing* data during World War II and now they needed a single library that contained information available to the entire industry. In a paper delivered at the 1980 fall meeting of the Rubber Division by Ruth Murray, an "information specialist" (librarian) of the division, she told the members, "Unofficially, the seed for the Rubber Division library was planted as long ago as 1920 by Miss Josephine Cushman, librarian at the University of Akron, when she published a small pamphlet entitled, 'A Special Library for the Rubber Industry.'"

Cushman credited, among several individuals, Hezzelton E. Simmons, president of the University of Akron from 1933 to 1951, for supporting the idea of a special library for rubber chemistry and the Rubber Division. Simmons, who took over the responsibility of teaching rubber chemistry from Charles M. Knight in 1913, always had a spot in his large heart for the arts as well as his chosen profession, the sciences. During his tenure, he helped foster relationships with industrial partners to create scholarships and fellowships at the University of Akron. These resulted in the beginnings of the master's degree program in rubber chemistry.[3] Simmons also was secretary/treasurer of the Rubber Division from 1928 to 1934, and his support of efforts to establish a Rubber Division Library fit his vision.[4] But it wasn't until World War II had ended that there was any action on the recommendation, under the jurisdiction and nurture of Simmons, who was still president of the university. Similar to Benjamin Franklin's requisite, a need provided the motivation and subsequent accomplishment.

In September 1946, as the first of today's "baby boom" generation was entering the planet, Willis A. Gibbons was nearing the end of his second consecutive term as division chairman. Because of World War II, general meetings were sparse, and the executive committee conducted most of the division's business—usually by telephone, mail, and an occasional meeting. "We had a most extraordinary executive committee," Gibbons told division historian Herbert A. Endres years later.[5] "Due to the prudence of that committee, we had a substantial balance, but we needed to invest that money."

At a meeting of the executive committee in Chicago, the finance committee, through its chairman Ray P. Dinsmore of Goodyear, recommended purchasing a one-thousand-dollar U.S. savings bond and, among other things, investing in a Rubber Division library at the University of Akron.[6] The committee proposed that the library contain materials in the field of synthetic rubber made available by the government, reports of rubber missions to Germany, and books, periodicals, and reports provided by the division. Benjamin S. Garvey Jr.—of B. F. Goodrich and, later, Sharples Chemicals—was appointed chairman of a subcommittee, the library selection committee, to further study the proposal and report to the executive committee at the next meeting.[7] Garvey, who had many scientific accomplishments in his career, including inventing the "Garvey Dye" during World War II, was also an ardent proponent of the library's development.

The members of that first library selection committee approved by Garvey were Harold C. Tingey of U.S. Rubber; Leora Straka, Goodyear; Harry N. Stevens, B. F. Goodrich; Dorothy Hamlen, the University of Akron; Frank Kovacs, Seiberling; Phyllis Hamilton, Firestone; and Ralph Appleby of DuPont. Later, the committee added Lois Brock of

General; Doris Hall, Firestone; Fern Bloom, B. F. Goodrich; Hezzelton E. Simmons, the University of Akron; and Ralph S. Wolf of Columbia Southern Chemical.

One of the first recommendations of that committee in their first official meeting on March 8, 1947, was to compile a "Union List of Serials" and to purchase some foreign journals.[8] "This was essentially a list of magazines from each of the rubber companies," recalled Lois Brock, a member of that committee and at that time a librarian for General Tire in Akron, Ohio. "By having that Union List of Serials, it would help us with inter-library [company-to-company] loans. The Rubber Division saw a great need for this exchange of information." The Union List of Serials, an inventory of technical and business journals, was supplemental to the *Bibliography of Rubber Literature* the division had compiled since 1936.[9]

At the division's spring meeting on May 26, 1947, in Cleveland, the membership approved allocation of five hundred dollars to purchase the journals, sponsor the list of serials, and collect records of the wartime developments in rubber. At the fall meeting, the division further recommended sponsorship "of the development at the University of Akron of a complete library on rubber technology." In a thesis for his master's degree from Kent State University in 1955, Jack W. Neely wrote of that historic fall meeting, "The official establishment of the library at this meeting marked the beginning of efforts to obtain one of the most complete libraries on rubber, resins and plastics in the world."

The newly formed library committee then faced the challenge of compiling, "mimeographing," and publicizing the Union List of Serials. Garvey appointed Leora Straka as chairperson of the operating subcommittee whose primary function would be to supervise its establishment and operation.[10] Straka, however, who had earned her Ph.D. in chemistry from the University of Cincinnati, was employed at Goodyear as a full-time librarian and, as such, could only devote a certain amount of her time to this extracurricular activity. In addition, the division wanted to acquire a rare and valuable collection of rubber literature owned by David Spence, the first Charles Goodyear Medalist in 1941, from B. F. Goodrich and instrumental in the early development of accelerators and antioxidants.

The wheels of progress were spinning. The division had recognized the need to collect and retain all the valuable data and information obtained during the war years to take advantage of the work already completed. However, who would assume responsibility of such a plethora of information and where would it be stored?[11] It was in relationship to this concern that the library committee recommended to the executive committee that a special librarian be selected and an endowment fund be raised.[12]

With the help of volunteers, the initial Union List of Serials was completed, made available in 1948 to members and key industry participants, and issued to cooperating libraries in November 1949. The prototype included all the educational and informational assets of Firestone, General, B. F. Goodrich, Goodyear, U.S. Rubber, and the University of Akron libraries.[13] DuPont and Glenn L. Martin Company later joined the venture. Primarily because of the advancement of electronic databases, researching material for the *Bibliography of Rubber Literature* was discontinued in 1971, but the last printed version was copyrighted in 1977.

In 1951, as if to elicit tacit approval of the compilation of the Union List of Serials, a few of the suppliers, along with the Rubber Manufacturer's Association (RMA), donated five thousand dollars to assist in the hiring of a full-time librarian to advance the heretofore totally volunteer efforts. Cognizant of the new funding source, on June 1, 1951, the University of Akron hired Betty Jo Clinebell, one of its own chemistry graduates, as the first official librarian of the Rubber Division. Clinebell was paid by the university but had a reporting role to Ben Garvey, chairman of the division's library committee, and Leora Straka, chairperson of the operating subcommittee (for policy matters), as well as Dorothy Hamlen, librarian of Bierce Library at the University of Akron, who ultimately was responsible to President Norman P. Auburn and the board of directors (for administration).

Since a Science and Technology Library already existed at the University of Akron, it was logical that the Rubber Division Library would become an extension of its genre-related cousin and, as such, would remain a "virtual library" during its lifetime. The first space dedicated to the Rubber Division librarian occupied a corner of the first floor of

Bierce Library on the University of Akron campus. It contained a modest 1,770 square feet of space and 405 feet of storage space in an adjacent department of the library building. The Rubber Division supplied the librarian two four-drawer filing cabinets and a typewriter.

While significant progress was made in a relatively short period of time, the division almost took two giant steps backward in 1953. That year, the University of Akron made a proposal to split the costs of the librarian, and as a result, the division's executive board, in a meeting on September 8, voted to discontinue sponsorship of the library by March 1, 1954.[14] The library committee met on October 30 to discuss the executive board's reaction, and President Auburn decided to attend that meeting. The committee naturally agreed that the library should continue and asked the executive committee to reconsider its initial decision. The committee, under the guardianship of Garvey, also approached the RMA again. On January 22, 1954, the division's steering committee met in New York City and voted to recommend to the executive committee that they keep the library providing that additional financing could be found before March 1, 1955. Of course, the executive committee agreed, and later that year the RMA contributed the necessary funds to keep the venture alive — for another year.

A period of uncertainty existed after the committees' volleys, and there was a decrease in the library's services attributed to a lack of sufficient publicity.[15] The motivation is uncertain, but Clinebell, the first librarian, resigned in August 1954 and got married. Pauline Franks, who held the title of university librarian from 1980 to 1983, when she retired from the University of Akron after thirty-four years' service recalled, "We enjoyed working with the Rubber Division because they subscribed to many journals. They were very cooperative especially during the transition when Betty Jo became their first librarian."

Clinebell was replaced by Lillian Cook (née Sutter). During the next fifteen years, the division would have six more librarians: Patricia Dreyfuss (1958–60), Sandra Gates (1960–62), Dorothy Hamlen (1962–63), Mary Sack (1963–64), Carrie Franks (1964–68), and Virginia Allenson (1968–70). Then Ruth C. Murray took the job in 1970, technically defined at the time as a "literature chemist." Murray tutored under the legendary Straka at Goodyear, as did previous Rubber Division librarian

Carrie Franks. Murray also gave credit to fellow Goodyear research library colleagues Judy Hale and Eileen Ambelang. To add to her credentials, Murray had graduated from Chatham College in Pittsburgh in 1944 with an undergraduate degree in chemistry. After Chatham, she worked in the research libraries of General Aniline (previously the Ansco film company) in Easton, Pennsylvania, and at Alcoa in Pittsburgh. Logic dictated that Murray would be a perfect fit.

Murray remembers her first visit to her new quarters. "I was in the converted Xerox room," she said. "And it was about as big as a table. The room was in the front of the old library . . . no space whatsoever. It was really undignified."[16] (The Rubber Division Library had been a part of the Science and Technology Library and in September 1958 was moved to the Knight Chemical Laboratory, "where it will be convenient for the faculty, students and research personnel who work in this professional area.")[17] Her reaction to the initial accommodations aside, she enjoyed her time as librarian of the Rubber Division—fully twenty-three years when she retired in 1993. "Everybody knew everybody. It was a very homogenous group, and it was a very great group to belong to. . . . I liked it very much."

Truly dedicated librarians love their work. They are often unrecognized and, in these times of massive electronic and online databases, are frequently taken for granted. Having already spent twenty-plus years in the business, Murray knew she had to continually promote the many services that the Rubber Division library offered. After all, she was officially an information specialist. In the world of academia, one has to have a master's degree in library science before he or she can be ordained as a librarian. That never seemed to bother Murray because she wore several other unofficial hats. Two of those were chapeaus of public relations practitioner and salesperson. Murray would sometimes travel to the local rubber groups, often with her counterparts at the tire companies—Straka, Fern Bloom of B. F. Goodrich, Phyllis Hamilton of Firestone, and Lois Brock of General.

"Every once in a while I visited those areas to promote our services," Murray said. "My message was that we have this library and we have so much literature here and we hope you will use it. It's part of your membership fee, so call me and I can look up something for you and send you that information. Some of the people in those groups never got to Akron and didn't know about our services. After all, they were paying me this huge salary, why not take advantage of it?" she chuckled.

Lois Brock traveled with Murray occasionally and was one of the founding daughters of the tire company librarians' sorority. "I remember driving with Ruth to a [division] seminar in Detroit," Brock said. "We were singing girl scout campfire songs all the way up there."[18] Brock received her undergraduate and graduate degrees at Kent State University in the classics. "My big opus was titled *The Influence of the Greek Precedents on Book Six of the Aeneid on the Sack of Troy.* . . . But it wasn't going to help me much in the rubber industry," she said, laughing. In 1946, she took a job at General Tire and organized their library. In the meantime, she decided to become more familiar with her new industry and enrolled at the University of Akron "to fill in my background in chemistry." Pertinent to her industry and her quest of knowledge, she couldn't have made a better selection. One of her professors was G. Stafford Whitby, the premier rubber chemist.

"Dr. Whitby was like a little bird," she fondly recalled. "He was very knowledgeable and I felt honored just taking some of his classes." Brock absorbed subjects in natural and synthetic rubber and even algebra. "I got my first 'C' in my college career in algebra," she said. "It wasn't from Dr. Whitby, but the professor . . . I can't remember his name . . . said my problem was that I was 'sub-vocalizing' or saying the process out loud. I never knew that was a problem."

Brock worked at General Tire for thirty-five years and credits Gil Swart, who had responsibility of the research department, for the existence of General's library. "He was very supportive of our work and he was the one who realized the need for a library." The other person Brock relied on during her tenure at General and her subsequent involvement with the Rubber Division was Leora Straka, manager of the Goodyear library. Straka was another of the tire companies' librarians, and she was admired and respected not only by her colleagues but also by professionals outside of her discipline. In 1965, Straka and Harold J. "Judd" Osterhof, director of research and development at Goodyear "were the moving forces behind the founding of the Center for Information Services, or CIS," said Brock. "I believe [the CIS] performed a valuable service at the time. They did a terrific job." Oster-hof was the division's Charles Goodyear Medal Award winner in 1971.

The CIS, based at the University of Akron, was a consortium of companies in the rubber and plastics industries and the University of Akron under the auspices of the Rubber Division. The CIS was intended to serve as a comprehensive abstracting and indexing information service for the rubber and plastics industry, covering not just chemistry, but also formulations, physical and mechanical processing and industry news.[19] In addition, the CIS was challenged with developing an advanced coding system for polymers, for which Sebastian Kanakkanatt was responsible.[20] It also developed a sophisticated computer program for searching the index terms assigned to each reference, which the university's computer department, under the direction of Robert Hathaway, completed in 1966. In addition to the computerized search capability, the center also published two biweekly publications: a bulletin of abstracts of technical literature titled *Polymer Literature Abstracts* and a bulletin of industry news titled *Polymer Industry News*. Under Director Panos Kokoropoulos, the center operated from 1965 to 1969, at which time Chemical Abstracts Service assumed some of its activities.

The abstracting/indexing staff included Eleanor Aggarwal, Eileen Ambelang, Joyce E. Brown, Ed Coolman, Diana Danko, Magda Abdel Latif, Ed Smith, and J. I. Smith. The CIS also took responsibility, in conjunction with the publishers of *Rubber Age*, for publishing the 1961–62 and 1963–64 editions of the *Bibliography of Rubber Literature*. Mel Lerner of *Rubber Age* was the executive editor, and Deanna I. Morrow of CIS was the assistant editor during those years. Straka and her counterparts from the other tire companies served as unofficial advisors to the CIS.

Straka continued to work with the division, especially with her tire company colleagues. "I relied on Leora to help me," said Brock. "Since our companies were so close together [General was within a few blocks of Goodyear in east Akron], it was within walking distance and I'd go over to ask her for advice. I first became active with the division through the Akron Rubber Group. I went in on Leora's coattails. If she suggested we go to a meeting, well we went, but we were pretty crafty. We went to [Los Angeles] one year for the Rubber Division then managed to attend a library convention in Hawaii. Of course, we always picked the more glamorous of the two." Straka and Brock's Rubber Division mentor, in more than just professional matters, was Ben Garvey. "He taught us how to attend conventions," said Brock. "He said to start at the top floor [of a show] and walk down, floor by floor, and get all the goodies as you go."

But one day after the two tire giants, Goodyear and General, were involved in a legal battle involving oil-extended rubber, Straka met Brock outside their offices and "she was crying," said Brock. "She had tears in her eyes and was distraught. She said, 'they won't allow me to have you over any more.'" A fifty-year member of the American Chemical Society, Straka retired from Goodyear research in the 1960s after thirty-seven years as a chemical research librarian in the information center. She was active in the Akron Section of the ACS as chairman in 1962 and councilor from 1971 to 1973. She died on January 10, 2001, at the age of ninety.

In 1976, the Rubber Division's library in the Norman P. Auburn Science Building on the University of Akron campus was renamed John H. Gifford Memorial Library and Information Center in tribute to Gifford, division chairman in 1971, who died in a tragic automobile accident in Texas on February 14, 1975. "He was well liked by everyone who knew him," said Benjamin Kastein, division chairman in 1975. Gifford devoted forty years of his life to the rubber industry. He served with B. F. Goodrich in tire compounding and curing and plant processing for eighteen years. He joined Witco Chemical Company in 1954 as technical service director and, in 1961, moved to Houston, Texas, to work with Continental Carbon Company. Kastein added, "It was an appropriate gesture on the division's part to honor the man who was not that far removed from the chairman's post when he passed away."

When Ruth Murray retired in 1993, it was going to be difficult to replace someone with her vast experience and years of service. Ruth had elevated the standards for the division's librarians. Then Joan C. Long added another element to the compound. Nevertheless, a familiar characteristic was passed from Murray to Long: the focus on chemistry.

Like many individuals in these times, Joan Long had retired once but wasn't really ready to stop working. After thirty-two years at the technical center of Union Carbide in Parma, Ohio, Long retired in 1992 because the company's graphite electrodes division was experiencing a downsizing. Equipped with three college degrees—a B.A. in chemistry from MacMurray College in Jacksonville, Illinois; an M.B.A. from Baldwin-Wallace College in Berea, Ohio; and an M.S. in library science from Case Western Reserve University in Cleveland—Long drove to Akron to apply for a job at the university.

She couldn't miss. She didn't miss. After all, even her mother, Helen Carlson, had been a librarian—first at LeTorneau Technical Institute in Longview, Texas, then at McDonnell-Douglas Corporation in St. Louis, Missouri—for more than a quarter of a century. In September 1993, she was hired as the division's tenth librarian. "I knew early on that this was what I wanted to pursue," Long said. "I got it from my mother. She gave me an insight into that profession."[21] By the time Long took over from Murray, the division had changed in many ways; yet ironically, it continued to uphold the original objectives: to establish a collection of source materials in the fields of rubber, plastics, and polymers and to ensure that the user receives the materials he or she requires.

During Long's tenure, the University of Akron and the Rubber Division successfully lobbied the American Chemical Society to recognize the contributions made by many members of the ACS, the Rubber Division, the University of Akron, and several key Akron-based companies to the Synthetic Rubber Program from 1939 to 1945. Long was a member of the Synthetic Rubber Historic Landmark Committee co-chaired by Charles Rader, former Rubber Division chairman, and Roger A. Crawford, president of the Akron Section of the ACS. The other members of the committee included Frank N. Kelley at the University of Akron; Lu Ann Blazeff of the Rubber Division; Robert J. Fawcett from B. F. Goodrich; Benjamin Kastein, a Firestone retiree and former Rubber Division chairman (1975); Anoop Krishen of Goodyear; R. Kent Marsden from the University of Akron; John D. Rensel of Bridgestone-Firestone; Cheryl L. Urban, the University of Akron; Andrew J. Walker, the University of Akron; Shelby J. Washko, Polymerics and Rubber Division historian; and Kristine L. Weigel of the Rubber Division. On August 29, 1998, their efforts culminated in the placing of a plaque near the Goodyear Polymer Center in recognition of the entire community's contributions as a National Historic Chemical Landmark.

After eight years as librarian of the Rubber Division, Long finally achieved real retirement on December 31, 2001. She spends some of her spare time these days serving on the board of trustees for her alma mater, MacMurray College.

In February of 2002, Christopher J. Laursen became the eleventh librarian of the Rubber Division of the American Chemical Society and in doing so broke with tradition: He was the first male to occupy that position[22] in the division's century-old history. However, like most of his predecessors, in addition to his library skills and his master's degree in library science from Kent State University, he had a solid background in chemistry. That science foundation was inculcated at the University of Akron, where he received his B.S. in chemistry. His affection with chemistry, though, was only peripheral. "In the late 1990s, I decided that being a chemist just wasn't for me," he said. "I enjoyed research, but [didn't enjoy] working in a laboratory around nasty chemicals."[23]

Disdaining the world of test tubes, Bunsen burners, and beakers, Laursen pocketed his undergraduate degree and "started investigating things and read an article how librarians were organizing information on the Internet . . . and that the Internet/World Wide Web was the next big thing," he said. "I thought I could do that, and I wanted to get involved in something related to computers and online databases, plus research. So being a librarian fit all those criteria." However, he had to take a side trip and a job with an executive search firm in Beachwood, Ohio. "In the early 2000s, the venture capital market went south and my company started laying off people," Laursen said. "So I started looking for a new job in late 2001." That's when he discovered an opening at the Science and Technology Library at the University of Akron. "I thought it would be a great fit for my chemistry and library background." His intuition and perception belied his youth.

"Chris brought to the university and the division the best of both worlds," said Jo Ann Calzonetti, head of the Science and Technology Library and one of two people to whom Laursen reports (the other is Ed Miller, Rubber Division executive director). "He has excellent chemistry and business knowledge. In addition to his professional skills and work ethic, Chris has a good personal character."[24] While many worthwhile contributions had been made by his predecessors to the development of the division's library services, Laursen elevated the research capabilities to a previously uncharted dimension.

"Technologically speaking, from a research standpoint, [the division] wasn't even up to date with the 1990s," said Calzonetti. "But with Ed Miller and Chris, they saw where they needed to go. When Chris took the job, the division didn't have all of their publications registered and they were losing out on some royalties. But he took care of that and got in touch with the Copyright Clearance Center to protect their intellectual property. He was also instrumental in getting papers indexed on SciFinder and even worked with the division's business manager to install automated billing and credit card purchases of scanned papers that they were now offering."

Laursen's other reporting line, Ed Miller at the Rubber Division, is equally pleased with the librarian's contributions. "Chris brought an unprecedented combination of skills to this position," said Miller. "Having degrees and experience in both chemistry and library science along with his personality, positive thinking and outstanding dedication to the rubber industry, has served to move the division forward over the past seven years in our service to members and their supporting organizations."

Laursen shares his superiors' viewpoints and vision and believes plenty of additional opportunities remain. "So much information has moved online and to the web that the job of the librarian has changed significantly," he said. "In the past, the division librarians spent much of their time researching information in books and reading printed journals to compile information for clients. I am now able to do most of my literature searching online, and my first starting point is often an electronic database."

Research occupies a great deal of his time, and in spite of the advancement of technology, many industries remain at a lava-crawl stage. "The rubber industry is an old and mature industry, so a lot of excellent knowledge on rubber science and technology still resides in printed sources," said Laursen. "Overall, gaining knowledge and becoming an expert in the scientific literature . . . that's one skill that you aren't taught in school and comes only from years of experience. In some ways I'm like the wise old man on the mountain. Most people are pretty Internet savvy these days . . . the problem is they don't know where to begin their search and that's where I can help."

Mark Bowles, founder of Belle History in Cuyahoga Falls, Ohio, and author of the Polymer Science fiftieth anniversary book, *Chains of Opportunity*, has taken advantage of Laursen's services. "Chris is a tremendous resource," said Bowles. "I can ask him the most obscure question and five minutes later he'll send me a PDF of the information and documents."[25]

With today's bombardment of technology and the multiplicity of sources available to researchers, like most endeavors in life that are condensed to their true roots, the heart of all activity is an individual—a live person, a human. Librarians go beyond the URLs, the HTTPs, and the HTMLs, and Rubber Division librarians, in particular, regardless of the era in which they served, have always responded in that manner. "What I enjoy most about my job is that it's something new every day," said Laursen. "Almost every research project is something different. Searching is a somewhat organic process, and you learn a great deal, yourself, as the research progresses."

THE ⑬ COUNTDOWN

*Rascals are always sociable. . . . And the chief
sign that a man has any nobility in his
character is the little pleasure he takes in
others' company.*

ARTHUR SCHOPENHAUER, 1788–1860,
GERMAN PHILOSOPHER

It was 1932 and the stars were likely arranged in some
bizarre astrological pattern that twinkled a premonition of
unusual events. Consider the following: In the first few
months of 1932, the British arrested and incarcerated
Mohandas Gandhi, Pope Pius XI met with Italian dictator
Benito Mussolini in the Vatican, the child of Charles Lind-
bergh was kidnapped, the Graf Zeppelin began a regular
route to South America, U.S. President Herbert Hoover

announced his support of arms limitations, and Paul von Hindenburg was elected president of Germany. Later in the year, the Winter Olympics began in Lake Placid, New York, the Revenue Act of 1932 was enacted creating the first gasoline tax in the United States at one cent a gallon, Franklin D. Roosevelt defeated Hoover in the November presidential election, and Hindenburg began negotiations with Adolf Hitler about the formation of a new government in Germany.

Meanwhile, during that year there were a few ideas "brewing" in the Rubber Division. Melvin E. Lerner was the new editor of New York-based *Rubber Age*, one of the trade publications that followed the rubber industry, documenting every move it made, both inside and outside of laboratories, research centers, and offices. "My first exposure to the Rubber Division was at a meeting in St. Louis, which was in 1932," he recalled in an interview with division historian Benjamin Kastein.[1] "I was a young fellow and it was a most unusual opportunity for me, but I had been told that the division had a reputation for being 'wild and woolly.' I quickly learned that it was not an exaggeration."

Over the years, an interpretation of Lerner's analysis, taken out of context, might suggest that the division was somewhat of a rudderless ship. That was never the case, and it is certainly not the case today. It does not mean, however, that some of the members' actions in that era were not flamboyant or colorful. The adage around that time, "Work hard, play hard," was certainly true in most industries and especially at occasions where a large group of men gathered outside of their normal work environments to conduct business and socialize. To a much lesser extent, the same is true today. It is often referred to as golf.

"Wild and woolly" aside, the 1932 meeting had particular significance in its role as a catalyst for what evolved into two defining yet very important social elements of the Rubber Division's history—and don't be misled, members did socialize. "The suppliers' cooperative cocktail committee came into existence at this meeting," Lerner told Kastein, "and it started when a group of major suppliers got together to form this committee. The spark plug of the committee was Al Brandt of B. F. Goodrich Chemical."[2] Allyn I. Brandt and his fellow committee members envisioned a method of occupying the time of some individuals who often exceeded the standard professional decorum that was expected of

147

division members. "They appointed people to their committee whose companies [employed the individuals concerned] ... and suggested that a little more modicum of decorum be maintained at future meetings. It worked like a charm."

The suppliers' cooperative cocktail had a noteworthy and dual role in the division: It gave the members, who were often competitors, the opportunity to meet and talk about anything in a "neutral" environment and without fear of verbal reprisal from their superiors. In addition, the original suppliers' cocktail preceded the annual banquet, and its mere existence encouraged attendance, which was one of the division's main objectives for the meetings, and since suppliers picked up the tab, as Lerner said, it indeed worked like a charm.

Lerner was another one of the legends who embraced the Rubber Division and his "task" with reciprocal gusto. He was one of the first "media members" of the division and, because of his title as editor (and later publisher) of *Rubber Age*, was a perfect fit to chair the suppliers' cooperative cocktail committee. After all, in his rubber company-neutral position, he could also solicit many sponsors for the reception, especially when funding was needed. Lerner also realized that his social and sales skills wouldn't hurt him either in pursuing suppliers as potential advertisers for his magazine. Still, he assumed the role with great delight and unofficially handled those responsibilities for three-plus decades until his retirement in 1973, when he received the division's first Distinguished Service Award.

In the interim, by 1947, the suppliers' cooperative cocktail committee had proven its substance, gathered momentum, and was about to add another cause célèbre to the socially significant value chain—a logical extension of its initial objective, the 25-Year Club.

To a first-time attendee, it looks like the introduction of contestants at an elderly statesman's beauty pageant. The emcee at the podium asks all the men to stand up, and then he begins to speak in an almost prayer-like yet declarative tone, "25! Stand!" Attendees usually are seated at several tables in a restaurant, dining hall, or large meeting room. Then, in melodramatic increments of one, the emcee continues his verbal volleys right up the numerical ladder: "26!" Individuals scattered throughout the room start to sit back down into their semicomfortable chairs. Just like in the Super Bowl or World Series, the last one standing is the winner. It's called the "Countdown," and it is one of the more revered activities in the history of the Rubber Division of the American Chemical Society. It is the defining ritual of the 25-Year Club.

Most clubs and even some of the most unstructured organizations have their roots in some semblance of a charter or bylaws or even a few words hastily scribbled in a smoke-filled room on a cocktail napkin. However, this wasn't the case for the early years of the 25-Year Club's luncheon, which was founded in 1948 outside of the structure of the Rubber Division. "It started because some manufacturers wanted to get together for lunch outside of the division meetings just to talk," said Ralph Graff, the recognized "father" of the sixty-year-old club.[3]

The beginning of the club/luncheon/fraternity rests, like most things in the division, within a committee. In the December 1947 edition of Mel Lerner's *Rubber Age* magazine, he wrote that "it has been proposed that a twenty-five-year club be established . . . and a committee has been named to formulate plans for such a club. Members of the committee are Dr. Herbert A. Winkelmann (Dryden Rubber), chairman; W. W. Vogt (Goodyear) and Simon Collier (Johns-Manville)." The following year, reporting on the fifty-third meeting of the division held at the Book-Cadillac Hotel in Detroit, Lerner wrote in the December 1948 issue of *Rubber Age*, "A special luncheon meeting of the 25-Year Club was held at noon on November 8 with William G. Nelson (U.S. Rubber) presiding as chairman and with 108 members in attendance."

Lerner further reported that brief addresses were made by Nelson, Winkelmann (called one of the prime exponents of the club), and Harry Outcault. Nelson also introduced the members with "longtime" tenure—Bill Higgins (1902), Charley Haynes (1904), H. Fuller (1904), and Ben Henderson (1906). Bruce Silver of New Jersey Zinc, who was chairman of the Eligibility Committee, reported that the committee had decided that membership in the club "should be open to anyone connected with the rubber industry for at least twenty-five years, but should currently be a member of the ACS (American Chemical Society) or the Rubber Division." The requirement of membership in the ACS explained the recognition of the club's initial "veterans," who actually preceded the date of the division's existence, 1909.

In that Detroit meeting Norman A. Shepard, of American Cyanamid and chairman of the Name Committee, reported that the committee favored the name "25-Year Club, Rubber Division, ACS." Reports of both the eligibility and name committees were approved, and the next meeting of the club would be held during the division meeting in Boston in May 1949, with John Bierer of Boston Woven Hose as chairman.

Since Lerner was intimately involved with the suppliers' cooperative cocktail, he was likely the catalyst that brought that endeavor together with the 25-Year Club luncheon. "He was a wheeler-dealer," Graff recalled. "He was a good friend to the division, but he had ulterior motives, too, to form these things."

The logic, indeed, didn't escape the crafty mind of Lerner. "I was invited to participate in the formation of these various groups . . . such as the suppliers' [cooperative cocktail] committee and I headed that committee for a period of almost twenty-five years. At the same time, I was one of the sponsors of the 25-Year Club—not because of what I am but because of who I was. I was the editor of a rubber journal. I was in a position to publicize and I was in a position to editorialize." He was also in a position to collect advertising dollars. Regardless of the personal benefits he reaped in his positions, Lerner "married" the suppliers' cooperative cocktail to appropriate sponsors and opportunistically joined the cocktail as the prelude to the 25-Year Club luncheon. "It was a nice fit," said Graff. "We eventually received excellent support from companies like DuPont, Cabot, Columbian, American Cyanamid, R. T. Vanderbilt, Harwick, Monsanto, United Carbon, and Binney & Smith. Cabot in particular was a great supporter."

From its inception, the 25-Year Club luncheon was not an officially sanctioned event of the division. It wasn't until 2003 that the group even had bylaws. While Herbert A. Winkelmann was the first recognized chair of the committee that formed the group, that position usually changed every year and it was a standard operating procedure for the chairman to also serve as the event's emcee. It remained that way for the next sixteen years until 1964, when Jim D'Ianni, chairman of the division, and the mild-mannered son of an Italian immigrant, became the first long-term chairman of the 25-Year Club.

When veteran members of both the Rubber Division and the American Chemical Society hear the name Jim D'Ianni, it's repeated with polite respect and admiration for his many accomplishments. Under the auspices of the Rubber Division alone, D'Ianni's achievements are numerous, primarily because he always was cognizant of the impact of seemingly small events to the overall objectives of the larger entity. He and his philosophy fit perfectly with the 25-Year Club.

While a relatively small part of the division's semiannual meetings, the 25-Year Club nonetheless has played a large role in fostering and supporting the fellowship that its founders envisioned in the first place. It was also important to the continuous interaction between manufacturers and suppliers.

"I always knew suppliers were very important," D'Ianni said in his final interview in June 2007:

> As much as anything else, they provided financial support where [the division] needed it, and the suppliers supported most of the division's major programs. The rubber companies were not in a very good position to be hosts and spend money on their customers. . . . It's just the opposite. The rubber companies were the guests of the suppliers, and so they benefit from that. And it's pretty much that way now. Even more so because if you look at the programs they have now, nearly all of them are often run by the suppliers.
>
> The cocktails before the banquet and 25-Year Club luncheon are the direct result of suppliers' contributions. They are very important to the division. We have been able to establish many solid business relationships with people like DuPont and Cabot and a lot of other suppliers. I got to know many of them through my involvement with the 25-Year Club.[4]

While D'Ianni was the recognized "founder" of the organization, the "father" of the 25-Year Club is Ralph Graff, who took the chairmanship in 1981 as D'Ianni became past-president of the American Chemical Society.

When World War II was over and the division, the rubber industry, and the world were getting back to more pleasant things upon which to focus their energies, young Ralph Graff had just completed his military tour of duty, acquired a B.S. in chemical engineering from the University of Pittsburgh in thirty-one months, and was working at the Goodyear Tire & Rubber Company on the production squad for the amazing wage of ninety-six cents an hour. "You could get three cents more an hour if you worked a shift between 6:00 p.m. to 6:00 a.m.," he said.

At that time of his life he was interested in just hanging on to his job and making those proverbial ends meet. While he was at Goodyear he admits that he went to a Rubber Division meeting in Cleveland only because Cabot, a supplier, invited him to go. In several capacities, Graff continued his relationship with Cabot and other suppliers for the next sixty years. Like many members of the division, Graff got his real introduction to the professional societies of the rubber industry via the route of the local rubber groups where he learned how to "work a room."

"I joined the Philadelphia Rubber Group in 1952 when I was working for DuPont," Graff said. "I became part of the banquet committee for the 1955 Philadelphia Rubber Division meeting and the chairman in 1980." In 1953, Seward Byam, a colleague at DuPont and division chairman, asked him if he was interested in joining the Rubber Division. Graff said he was also encouraged to join by George Vacca of Bell Telephone Laboratories and Arthur Neal, another DuPont colleague. Neal (1954–57) and Vacca (1964–66) served as secretaries of the division. From 1966 to 1968 Graff was a Philadelphia area director and, likely because Byam was advertising manager for many years, "in 1967 I was approached for advertising manager for the division's publications," he said. After three years, Graff became assistant treasurer (1970–72) and treasurer (1973–75) of the division. In 1979, the 25-Year Club became an official activity of the division. "This was about the time Al Laurence of R. T. Vanderbilt was retiring and D'Ianni and I convinced the executive committee, and it was approved," said Graff.

Albert E. "Al" Laurence of R. T. Vanderbilt was division chairman in 1972, but he had been involved with the 25-Year Club when Mel Lerner was peddling his advertising. In addition to his leadership skills, and because of his large size, Laurence was often referred to as "Big Al." In the mid-1970s, after an incident in a meeting in Cleveland where a would-be thief busted into division administrator Marge Bauer's room, hit her a few times, stole thirty dollars out of her purse, and then left, Big Al also was asked to wear the additional hat of head of security.

———————

The Rubber Division always has had a very strong relationship with its suppliers. To name one or just a few would be a disservice to the many others who were significant contributors—financially or otherwise—to the division's endeavors. The division veterans know the importance of the suppliers to the entire operation. Ralph Graff relayed the routine at early meetings involving suppliers: "The meetings started at nine o'clock in the morning and broke for lunch at eleven or eleven-thirty. Then the suppliers opened up their hospitality suites, and they had raw bars, and clams, and clambakes. . . . Then we opened up the meetings again at two o'clock and they went to four-thirty, and then people dressed for dinner. Then they had a pre-dinner cocktail hour prior to the banquet."

The custom in those days also included postmeeting evening activities in some supplier's hospitality suite. "After midnight there were usually some poker games started," Graff admits, "and it was also an unwritten rule that as long as there was somebody standing up, why, the supplier had to keep his suite open. In New York we used to close at midnight, get the crowd and go over to the Latin Quarter to the one o'clock show, come back to the hotel by 3:00 a.m., and then go out for breakfast." Of course, because of many concerns and legal implications today, those activities are just memories of "good old days." Graff added, "You could also gauge a company's success for one year by the size of the shrimp they offered in their hospitality suites."

In 2005, Graff handed over the chairman's duties of the 25-Year Club at the spring meeting in San Antonio to John R. Deputy of Americas International. He also gave Deputy the remaining memento pins that Cabot had donated for so long to the "last standee." Deputy joined the rubber industry in 1964 with B. F. Goodrich in their technical training group on South Main Street in Akron, and became eligible for the club in 1989. "Interestingly enough, I joined the industry (1964) the same year that Jim D'Ianni was chairman, and now, many years later, I have the high honor of being the chairman of the same club of which he was the first chair," Deputy said.[5]

Deputy is also a precise fit for the position. He loves his work. He loves people, and he loves the division. "As usual I spend most of my time doing what I love to do—caring for customers in the rubber industry," he said. "I love telling them about my company's products and how to use them to get technical performance—as well as value—at the same time from my products. Of course, I enjoy working with the division and my new role with the 25-Year Club." His vision for the future of the club might not be awe-inspiring or as profound as working on world peace, but given the structure of the organization, it is very appropriate. Deputy said, "We ran out of the Cabot pins a couple of years ago and Americas International supplies them now, but I have to order some new ones."

Besides the renowned Lerner, the social events (and, of course, the meetings) attracted other media moguls, the most notable of the modern era being Ernie Zielasko, a veteran tire and rubber industry journalist who received the division's Distinguished Service Award in 1985. While Lerner was a glad-hander and back-slapper, Zielasko was recognized for his professionalism. He was respected not only within the division but throughout the entire industry and started his career in the public relations department of B. F. Goodrich in Akron, then with *Rubber World* and *Modern Tire Dealer* magazines, and ultimately as founder of *Rubber & Plastics News*.

Zielasko became familiar with the Rubber Division when he was at *Rubber World*. "I remember I drove with our sales manager, Chris Chrisman, to a meeting in Washington, D.C.," Zielasko told Benjamin Kastein in 1986. "I was a neophyte and really didn't understand anything." But in 1971 he founded *Rubber & Plastics News* and, by that time, knew the industry well enough to buy booth space for the October meeting in Cleveland. "I got a real feel for what the division was by attending that meeting," he recalled. "I was so impressed with the dedication of all the people who were working there. While they were paid by their companies, they did not make one red cent from the division. I had been to several trade shows where they had paid staffs and I was more impressed with the way the division's volunteers worked. They worked just as efficiently."

Zielasko was particularly impressed with the work of the local arrangements committee. "They worked very hard," he said, "and I thought they ought to be recognized." That's when he decided that the idea of a booklet detailing the events of the meetings would be a worthwhile medium to (1) publicize the meeting itself and communicate the activities and (2) earn a dollar or two for his company in the process. The "FAX," also known as the "Yellow Book," was born in the fall of 1972 in Cincinnati. The five-by-seven-inch booklet contained everything that the later "Show Daily" offers, albeit in a greatly condensed version.

Ed Noga, editor of *RPN*, joined Zielasko's staff in 1979 as managing editor. He recalled his boss and "FAX." "It was always a sore point between us," he said. "I considered it public relations, and not newspaper journalism, so we'd argue over it. He agreed, but did it as a service to the Rubber Division members, and also to make a few bucks in ads."[6] Noga also became enamored with the division and its operations. "For many years we made a point of showing up at the 'countdown' of the 25-Year Club, and taking a picture of the winner (the person with the longest tenure in the industry who was at the lunch), and publishing it," he said. Then he added a humorous sidebar: "Our editorial staff, particularly when it was populated by twenty-somethings, used to affectionately call the 25-Year Club the 'Dinosaur Club.' Well, for about five years *I've* been eligible to be a member. I want to be a stegosaurus."

Rubber & Plastics News discontinued "FAX" and the subsequent "Show Daily" in the early 2000s, when the division had its changeover. However, on August 13, 1984, *RPN* published a special edition of its magazine titled *In Tribute to the Chemists Who Tame Rubber*, in recognition of the division's seventy-fifth anniversary. Benjamin Kastein, Ralph Graff, and Walt Warner also contributed editorial assistance to that special edition.

Noga continues to work toward his status as a representative of the modern day Jurassic period. Zielasko passed away in December 2003 at the age of eighty-four.

Another modern era media member who has been a frequent follower of division activities is Don Smith of *Rubber World*. *Rubber World* has been covering the rubber industry longer than any publication in North America, beginning in 1889 as *India Rubber World*. In 1966, the magazine issued the first of a five-part series on the division titled *The Division of Rubber Chemistry: Catalyst of an Industry*. The series was a very comprehensive account of the division's activities to that date, including "concrete terms which the hard-headed appreciate most: money, activities and personalities."[7] Its authors were Otto Scott, Bill Mulligan, Joseph Del Gatto, and Ken Allison.

In the magazine's Centennial Edition in October 1989, which devoted the majority of its content to division-related information, Smith wrote in part, in his opening editorial, "To read of the evolution of this industry on a month-to-month basis is remarkable, because it's not only an account of the progress of a material and an industry, but an account of extraordinary progress of the human species."

The media had more than just an auxiliary role in the division. There have been a few members that were employed by the fourth estate. They include 2001 chairman Rudy School (*Rubber & Plastics News*), who was a former full-time employee for that publication, and a few technical editors who wrote for several publications on a regular basis. Those writers were Benjamin Kastein and Walt Warner, and because technical writing was commonplace among many members, virtually hundreds of others.

All of the "unofficial" activities of the division were certainly an integral part of its structure and history, and the force behind them, the volunteers, were like extras in a Cecil B. DeMille movie. Zielasko observed, "I think there's a love affair of the division among many of the members. I think I became a victim of that, too."

PART SIX

MATURITY

METAMORPHOSIS

*Business is never so healthy as when, like a
chicken, it must do a certain amount of
scratching around for what it gets.*

HENRY FORD

In a fifteen-year period from the early 1980s to the mid-
1990s, the rubber industry had bounced right into that
proverbial brick — or hard rubber — wall. The tire industry,
the most prolific of elastomer consumers, was affected the
most. Consolidations, downsizings, layoffs, mergers, and
takeovers were the bywords of the era. From 1981 to 1992,
thirteen separate mergers or acquisitions took place within
the tire industry. Only two of those failed. One involved
U.S. giant Goodyear, but that company's fending off of Sir
James Goldsmith's forays in 1986 left it a much different

organization. The other failure was Pirelli's 1989 attempt to acquire Continental, A.G. The merger mania, improvements in technology, globalization, and volatile economies were felt down to the factory floor. The war years—especially World War II—notwithstanding, the U.S. tire industry's employment reached a peak of 123,700 people in 1975, but by 1995 that figure had fallen to 80,200.[1]

Not surprisingly, the Rubber Division of the American Chemical Society felt the large ripple effects of the industry's tidal wave of motion and adversity. At its height of activity in 1985,[2] the division had a membership of 5,081,[3] but by the beginning of the twenty-first century those numbers had dwindled to slightly more than 2,000. For many reasons, it came as no surprise to the division leadership that the organization had reached a critical point in its existence.

"After reviewing a financial report in the mid- to late 1990s, it was fairly evident that if we didn't make some changes we were soon going to be turning out the lights," said Rudy School, secretary in 1998–99, and a member of the division's steering committee who would become chairman in 2001, the second of two critical years—2000 was the other—in the division's history. "In 1988, I had presented to the board a letter in which I suggested [the division] move away from their current structure and have the group run by an executive director. Then, I applied for the job," he recalled.[4]

School's missive of 1988, and his subsequent recommendations, were of interest to the rest of the committee, but the group wasn't yet ready to make that kind of commitment. Several other activities were occupying their collective minds. It wasn't until a few years later that a motivation would materialize. In the meantime, the impetus to change—the motivation—began to take shape, but there were two issues that had to be resolved. One was School's recommendation that the division alter its internal structure, and the other was strictly physical. Before they could attract that full-time director and a complementing staff, they needed additional office space to accommodate a proposed new internal structure.

Ever since April 1, 1991, when the university opened and dedicated its new Goodyear Polymer Center, the Rubber Division had operated its four-hundred-square-foot offices there. Prior to that, Harwick Chemical (July 1963–September 1963), Knight Hall (the old Knight building,

1963–65), Leigh Hall (at that time often generically referred to as the Business and Law building, 1965–68), the Norman Paul Auburn Science and Engineering Center (1968–76), and Whitby Hall (1976–91) served as offices for the division's first full-time, functional operations person. That was Marjorie Ella Bauer, essentially a one-person office.

Marge, as most everyone called her, was the consummate administrative assistant—referred to as a secretary, a very admirable and respected moniker in the pre–politically correct era. She came from a small but proud family. Her father, Frank Uhl Bowen, moved to Akron from the state of Washington in 1918, owned a grocery store, worked in real estate, and then worked with a construction company before landing a job with the Daniel Guggenheim Airship Institute near the Akron Municipal Airport as a night watchman, from which he eventually retired. Frank married Anna Elizabeth Hickman, and in 1916 the couple had their first of two daughters, Grace. Four years later, the Bowens were the parents of another girl. Marjorie Ella was born on December 28, 1920.

Marjorie grew up in Akron and went to Garfield High School, where she graduated in 1938. She was very proud of the certificate she earned in 1939 from Akron's Hammel Business College on East Market Street in their complete secretarial science curriculum. Then Bauer took her first job at Foster Office Supply on Bowery Street. She received a weekly salary of twelve dollars for her forty-six-hour assignment but stayed there only five months before taking a higher-paying job at Pittsburgh Plate Glass in Barberton, where she earned a monthly salary of seventy-five dollars for only forty hours a week. On December 12, 1941, she married Daniel Ralph Bauer, and shortly thereafter Daniel was drafted into the U.S. Army—a common occurrence for young men at the time, especially after Pearl Harbor.

In July 1942, Daniel was shipped to Biloxi, Mississippi, and Marge went with him. She didn't know if she'd be able to see him again, so she took a job at the base, Keesler Field, and stayed there until August 1945, when he was released and they moved back to Akron. Marge gravitated to Guggenheim, where her father had worked, and in September of 1945 she took a job there as a secretary. She left Guggenheim in December of 1947 because she was expecting her daughter, Pamela Diane, who was born on April 5, 1948. After working in her sister's flower business for a brief period, she took a job at Goodyear Aircraft as a secretary to the sales manager, Denton Zesiger.

In May 1956, she started working for Milton Leonard, technical manager at Columbian Chemical, a major supplier of carbon black to the rubber industry. Leonard also was an active member of the Akron Rubber Group. He was secretary, treasurer, and, eventually, chairman of the group. "Up until that time, I didn't know there was such a thing as carbon black," Bauer recalled.[5] Her time at Columbian was brief because the Bauers decided to move to Florida. However, a seed was planted at Columbian by Leonard that would bear fruit for Bauer when she would return to Akron in late 1962.

"Just before I left for Florida, Mr. Leonard, who was on the executive committee of the Rubber Division, wrote a letter to the officers saying that the Rubber Division really needed an executive secretary to handle some of their administrative work. At the time, I didn't know anything about the Rubber Division." Bauer was gone for just a few months, but during that time Leonard teamed with fellow Columbian colleague John W. Snyder to begin writing a chapter in Maurice Morton's book, *Introduction to Rubber Technology*,[6] dealing with carbon black. Bauer later said that she helped to type some of Leonard and Snyder's manuscript.[7]

In the meantime, at a steering committee meeting on November 9, 1961, Rubber Division chairman George E. Popp began soliciting names of members to form a committee to evaluate the need for an executive secretary-treasurer.[8] The following spring, on April 24, 1962, the executive committee approved the hiring of an administrative secretary and the establishment of a Division of Rubber Chemistry office for a trial period "of two years with the cost not to exceed ten thousand dollars per year."[9] The minutes further stated that the administrative secretary is to "take care of some of the routine jobs that are being done by the [division] treasurer and the secretary."

That summer, division treasurer Dale F. "Bud" Behney of Harwick Standard Chemical began investigating potential office space at the University of Akron. At the October meeting in Cleveland, Behney told the steering committee that he had found a three-hundred-square-foot office for the proposed administrative assistant, but the space wouldn't be available until 1964. It would cost only six thousand dollars.[10] The committee agreed and approved Behney's proposal.

In February 1963, at another steering committee meeting, this time in Toronto, Behney offered the group an alternative to an executive secretary. He had contacted a local management organization, the Association Administration and Services, Inc., operated by Thomas A. Bissell. At a follow-up meeting in May "no agreement was reached concerning Mr. Bissell's proposal to serve the Division." The vice chairman of the steering committee, Jim D'Ianni, suggested that the group consider Ralph Wolf for the position, now estimated to cost the division between thirty and thirty-five thousand dollars a year. While the committee just mulled several options, there was action on another front.

"When I got back [from Florida] in April of 1963," Bauer reflected, "I got a call from Milt Leonard saying that he was being transferred to New York, his secretary was leaving and he wanted to know if I could come in and work for a while until he moved. So I did." While Bauer was working for Leonard, he mentioned to her that there was a job at the University of Akron that might be a good fit for her. "So I went over [to the university] not realizing that I was going to be given a test," she said. "When I arrived there they gave me an administrative test. They asked me to start work on July 1 for the Rubber Division."

However, there was one small problem. The Rubber Division still didn't have any offices where Bauer could work. For more than fifty years, records, files, and other material for the organization had been the responsibility of each officer and his or her particular discipline. The treasurer kept the financial files, the secretary kept the correspondence, and so on. The university had to find at least a temporary office for Bauer and her new client. "Other than a building that was to be torn down, there was no space available on campus," Bauer said. The Rubber Division's treasurer, Behney, and the treasurer-elect, John H. Gifford of Witco Chemical, accompanied Bauer and some university personnel to the aging building. "When we couldn't get the door unlocked with the key, Bud Behney told me I could come out to Harwick and work [from there]," Bauer recalled. "So I reported to work [for the division] at Harwick."

The first full-time "employee" of the Rubber Division, Marge Bauer was technically a state of Ohio and University of Akron employee who was assigned to handle the office of the treasurer of the Rubber Division of the American Chemical Society from the offices of Harwick Chemical where she shared an office with salesman Gardner Brown. Fortunately for her, in mid-September, the university found her an office on campus in Knight Hall. "We were on the first floor, Room 14," Bauer said. "Bob Pett was working on his doctorate in the lab next to my office, and Dr. Morton used to tease me that he was saving me a place in the basement of the Auburn Science Building, but I told him I was already on the ground floor," Bauer said with a laugh.

Maurice Morton, director of the Institute of Rubber Research at the university, and Robert A. "Bob" Pett were two fellow building tenants of Bauer and the Rubber Division in Knight Hall. Pett would later serve as division chairman in 1988 and become the first, and to date the only, division officer who worked directly for the auto industry. He was one of the last two recipients of a doctoral degree in Polymer Chemistry from the University of Akron.[11] Pett retired from Ford Motor Company as their worldwide internal consultant on rubber on January 1, 2002.[12]

Morton, of course, is a legend at the University of Akron. Born in Latvia, Russia, in 1913, he and his parents immigrated to Canada in 1921, and by 1945 Morton had his master's and doctorate degrees from McGill University in Montreal. From 1936 to 1941, while he was completing his education, he was chief chemist for Johns Manville in Asbestos, Quebec, about one hundred miles northeast of Montreal. In 1948, George S. Whitby, who received his master's (1918) and doctorate (1920) from McGill and had traveled to Canada periodically as a consultant for efforts on the new synthetic rubber program, knew of Morton and summoned him to the University of Akron to become assistant director of Rubber Research. When Whitby retired in 1952, Morton took over the program, and among many other achievements in the field of rubber chemistry and polymer research, Morton later teamed with Howard L. Stephens, a colleague, in 1963 to develop the Rubber Division's first correspondence course. The preliminary work actually began in 1956.

In a recorded interview of Morton conducted in 1981 by division historian Benjamin Kastein, Morton reflected about his first book, *Introduction to Rubber Technology*, and the subsequent spin-offs. "The book itself was basically a compendium of lectures which had been given around the country by the various Rubber Groups, and, of course, all the Rubber Groups had done a fantastic job in furthering education in rubber technology," he told Kastein. Morton credited Arthur Juve, division chairman in 1956, with the idea of putting the best of those lectures into the book. In 1962, Morton's *multum in parvo* became the catalyst for another idea.

"Cap Lundberg was chairman of the [division's] education committee," said Stephens, who received his doctorate from Akron in 1960, "and one day he came to Morton and said that all of the material in the old textbooks are out of date. [Lundberg said,] 'How about updating your *Rubber Technology* book?'"[13] Morton liked Lundberg's suggestion. "[Lundberg] felt that it was time that the activities of the Rubber Groups—which had been simply to schedule a series of lectures in various cities—could remarkably be enhanced if we made that material available to the whole rubber industry regardless of whether they lived in the cities where they could attend lectures or not," Morton said. "And his idea, and the idea of the education committee was: Why not have a correspondence course based on my book? After all, the book now represented what people were getting in the lectures."

Morton and Stephens then collaborated on the pioneer correspondence courses in rubber technology that are still popular today. Stephens, who retired from teaching in 1983 and became a professor emeritus of polymer science, said the courses were almost an immediate success. "When the course was first offered, we had over six hundred applicants. We had to get graduate students to help us grade the papers so we decided to limit it to three hundred—but we split them up with three hundred in the spring and three hundred in the fall."

In 1981, Morton estimated that more than 10,800 students had completed the course. Today, while the standard six hundred applicants a year has declined to about two hundred a year, the course has retained most of its original 1960s base with slight revisions, including topics such as an introduction to polymer science, the compounding and vulcanization of rubber fillers: carbon black and non-black, processing and vulcanized tests, physical testing of vulcanizates, natural rubber, and styrene-butadiene rubbers. The only changes to the original subject offering were "periodic updates and, with the suggestion of Harold Herzlich, we put together a more advanced course," Stephens said. To complete the update, in 2004 the division put the courses online via their web site, www.rubber.org.

Like a true educator, Morton, who was chairman of the Polymer Division of the American Chemical Society in 1962, summed up his thoughts about his books and the resultant correspondence courses: "It's amazing. It really tickles me whenever I travel the United States or overseas. There are always autograph seekers," he laughed as he told Kastein. "They come up and say, 'Dr. Morton, I took your course. It was very interesting. Would you please sign my copy of the book?'"

Stephens remembers that Morton was very fair to him in the collaboration of the correspondence courses, "but he always let you know that he was the boss," he said. "I got along with him fine. I handled a lot of details for him, requisitions, etc. But we had some guys from the Air Force who used to work with us and they gave [Morton] a nickname: 'Maurice the Magnificent,'" he laughed.

In 1978, when Morton was forced by law to retire at age sixty-five, he selected one of his former students, Frank N. Kelley, to succeed him. Kelley recalled that Morton "was an intimidating presence. In spite of his bulldog countenance—he had a lot of chutzpah—he was a sympathetic person when I went to him with a problem. I had great respect for his accomplishments."[14] In his official eulogy, Kelley called Morton "an inspiring teacher and scientist who gained international respect in his field. Dr. Morton laid the foundation for Akron's academic polymer program, and his work is still benefitting the college, its alumni and the polymer industry."[15]

Morton died on March 23, 1994, of mesothelioma caused by exposure to asbestos. He was eighty-one years old.

Marge Bauer habitually referred to her office as "we," but admitted she was really a one-person operation. "We" worked out of Knight Hall for about eighteen months. "We had one wall-plug which handled an adding machine, a twelve-year-old typewriter and an air-conditioner. Other than Dr. Morton's office—and mine—the building was not air-conditioned," she boasted. In January 1965, Bauer and a recently added student assistant who worked for about fifteen hours a week moved to a new building, Leigh Hall. By 1968, Bauer and her part-time student assistants had moved to the Norman Paul Auburn Science Building. "That put us with the people at the University that we were working with," she said. "In the other building we had law professors, etc."

By the early 1970s, Marge began adding to her duties and acquired "some of the secretary's work, but not yet the chairman's," she said. She even received the okay from the university to add a full-time helper. In June 1972, almost ten years after she arrived on the scene, Marge hired Constance "Connie" Morrison. Morrison had recently graduated from Akron's North High School in 1971 and was eager to jump into the employment pool. "I had just been in the working world for six months when I started at the Rubber Division," Morrison said. "Marge became my teacher and mentor, and she was just so patient and giving of herself to me as I was learning the job."[16]

When Bauer began working for the division in 1963, membership was 1,388. By 1975, it had grown almost threefold to more than 4,255 and had added another part-time person. They needed more space. "In 1976, we moved to Whitby Hall," Bauer said. "This gave us the additional space we needed because we started doing all of our own mailings. This is when we began to gain momentum working with the secretaries, and doing all their mailings and meeting notices." For almost a decade and a half, the division offices remained in Whitby Hall, in a modest office location on the first floor of the building—just up the stairs and to the left as you entered the building—in a space liberally estimated at five hundred square feet. In that time, the division grew in many areas. Most notably, they entered the computer age.

"There was a song that we used to sing," said Morrison. "I don't know how I remember it, but the song was called the 'Wang Wang Blues,'" she said with a laugh. "We switched all the division information and mailing lists over to those Wangs." Instead of manifesting the real blues, Bauer and Morrison became the start of an uplifting, solid base for the Rubber Division. "I believe that patience had to be one of her strong characteristics in the world of the Rubber Division. She was a joy to work with and I was blessed the day that she hired me," Morrison said of her mentor.

Bauer retired in 1987. "She was the glue that held the Division together," remembered Pett. "The division truly couldn't have operated without her."[17] Morrison married Joseph Koons while she was working at the Rubber Division, and today she remains an employee of the University of Akron as the events coordinator for the School of Law. She said that her fondest memory about Marge Bauer was the special way she started conversations on the phone:

"Got a minute?"

Frank N. Kelley is as "Akron" as one could be. He was born in the Rubber City on January 19, 1935, and received his B.S. in chemistry from the University of Akron in 1958, an M.S. in 1959, and a Ph.D. in polymer chemistry in 1961. His first job while still in high school at Randolph (now Waterloo in Portage County) was at Akron's Goodyear Tire & Rubber Company on East Market Street. Then, after a series of assignments in several locations with the U.S. Air Force, first as an officer and then as a civilian primarily involved in scientific research, Kelley returned to his alma mater in September 1978 as the director of the Institute of Polymer Science. In 1988, when the university formed the College of Polymer Science and Polymer Engineering, Kelley was appointed dean of the new organization replacing the legendary Morton.

Kelley's job included responsibility for administration of the largest academic polymer program in the United States. In addition to the complete menu of undergraduate and graduate offerings in polymer science and engineering, Kelley was accountable for several research centers, a research lab, a training center, and administration of the staff of the Rubber Division of the American Chemical Society. "I knew of the Rubber Division when I was in high school in 1952 and working at Goodyear," Kelley said. So in 1978, when he returned to Akron and the university, he was very familiar with each entity. It was truly a homecoming by all measures for the man who had earned an Air Force Commendation Medal, two civilian Meritorious Service Medals, and the Exceptional Civilian Service Award from the secretary of the U.S. Air Force.

While the next twenty years were certainly instrumental in Kelley developing his impact on these components, a few of the events in his tenure during the 1990s would have a significant impact on the future of the Rubber Division. The catalysts for these measures were, strangely, new buildings.

The first of the new structures manifested itself in 1991, when the 146,000-square-foot new Polymer Center building was completed to house the Department of Polymer Science and the Institute of Polymer Science. The second manifestation came on October 21, 1997, when Rubber Division chairman J. Marshall Dean III publicly announced that the division would pledge one million dollars to the University of Akron and its Polymer Program, and subsequently, the Rubber Division would receive priority consideration for appropriate office space in the college's new building, which was to be constructed on campus by the advent of the new century. Although, since the Rubber Division was and continues to be a nonprofit organization, by all legal guidelines, the pledge could not be a donation but had to be applied to an arrangement — preferably long-term — for office space.

When Marge Bauer retired in 1987, her replacement as division coordinator was Lu Ann Blazeff, and the office support staff had grown to eight people. Whitby Hall, dedicated in 1973, and the present-day home to the School of Chemical and Biomolecular Engineering, could no longer accommodate the Rubber Division. "It was obvious that we had outgrown the office space in Whitby," said Connie Morrison, the first person hired by Bauer. "We had people on the first floor and offices also on the second floor."

The move to the new nearby Polymer Center in 1991 was the correct formula to cure the Division's working-space dilemma. It was at that time that the Rubber Division relocated its offices from Whitby Hall to the new glass-walled conjunction of two towers — one twelve-stories and the other nine-stories tall. The building also contained a 213-seat auditorium and a sufficient amount of space to house several campus tenants who were intent on growth. With the new building in place, the campus' new polymer towers were well on their way to establishing and solidifying their identity on the university landscape. There was still one "molecule" or two left, however, to complete a true bond with rubber.

"The offices in the Polymer [Center] were perfect," Morrison said. "Of course, everything was new — new furniture, desks, et cetera. So it was exciting to have new space and furniture. We got to have a say in what we got. I think that having everyone at close proximity lent to better communication between staff. 'It just worked better!' Having an area specifically designed as a 'mailroom' was a dream come true."

In 1991, using push carts and a variety of other moving devices, the Rubber Division staff trekked across a wide walkway to the north of Whitby Hall into the "spacious" thirteen hundred square feet of office space on the third floor of the Polymer Center Building, directly opposite the offices of the dean. The first inhabitants of the new offices included Blazeff (room 313), Morrison (319), Susan School (314), Marj Riccardi (320), Sherri Poorman (318), Kristine Weigel (317), Kayanne Toney (316), Vicki England Patton (reception), and student assistant Missy Beynon.

In the late 1990s, corporate donations seemed to be in vogue. On the heels of the Rubber Division's one-million-dollar gift to the university in 1997, the following year, one of Akron's largest employers, the Goodyear Tire & Rubber Company, thought it would be appropriate to donate first, three million dollars to the University of Akron to establish a Goodyear Chair of Intellectual Property in the School of Law and second, the Goodyear Global Scholarship Program in the College of Business Administration's International Business Program. Under the encouraging eye of chairman and CEO Robert E. Mercer, Goodyear kicked off the funding of the building with a "seed" gift of eight hundred thousand dollars and assisted in acquiring further support from several of its major suppliers. To reciprocate, the university renamed its seven-year-old polymer science building the Goodyear Polymer Center.

Many years later, in his retirement, Mercer still recognized the value of the chemist and research to his company and the tire industry. "An important source for our tire division came from our chemical division where many of the compounding materials and synthetic rubber [they developed] wound up in our production activities at a very competitive cost through transfer pricing, and quality was controlled to insure that their final product integrity was protected," he said.[18]

By the time the new millennium arrived, *change* already was a part of the new mantra. With Rudy School's suggestion of a full-time director and different structure planted in their minds, the executive committee also realized that a different direction was imminent. On January 1, 1999,

Chris Probasco, rubber chemical account manager for Flexsys, took over as treasurer of the division. He was assisted by Kurt Nygaard — executive vice president, sales and marketing, of Akron-based Harwick Standard Distribution — who became assistant treasurer. "Obviously, one of the first things we did was look at the 'books,'" Probasco said. "Our expenses far outweighed our assets, and it was like a train wreck!"[19]

That spring the division held its technical and committee meetings at the Palmer House in downtown Chicago. The Windy City gathering included the eleven members of the steering committee and twenty-six members of the executive committee, which included the area directors. "The division was having to dip into reserves just to pay bills," said Probasco. "Our revenues were shrinking, and we had to do something to change that." The fall meeting of 1999 in Orlando, Florida, gave some consideration to the financial state of affairs experienced by the division. "When we made a presentation to the executive and steering committees in Chicago earlier that year, many of the attendees were shocked," said Nygaard. "It was not common knowledge among the membership or its area directors."[20] One of the first things that group initiated was to schedule the expos and mini-expos in locations where they had large memberships — such as Akron, Cleveland, and Pittsburgh. These "trade shows" were one of the division's main sources of revenue, so in order to increase the potential of a larger turnout, they were logically planned for venues closer to members' homes or places of business.

Also in 1999, Rubber Division chairman Donald Mackey was informed that, if necessary, the University of Akron would be willing to allow the division to operate outside of the university system. The timing was both fortuitous and fortunate. "The University was willing to go either way," Mackey recalled. "We had several conversations about the subject, but nothing was decided until later."[21]

The Rubber Division was driven by a few key requirements: It needed to combat the decreasing membership and hence the decreasing revenue. In addition, the proposed change involved people. As the new century and fears of "Y2K" approached, the Rubber Division was focused on, among other things, the issues of solvency and its future. With the dwindling membership, limited sources of revenue, and a subsequent decline in numbers of the all-important volunteer pool, the division faced a critical turning point in its history. To add to the dilemma, as the new millennium approached, the ninety-year-old organization still did not have a traditional employee—paid directly by the organization—to oversee the business. Rudy School was a division visionary and realized as early as 1988 that the large ship was off course and headed for a virtual iceberg if it didn't change its direction. He knew the right course, but he couldn't tell the ship's crew *how* to steer it. School put the situation succinctly into perspective: "When you can't manage people, you can't manage the organization."

Trepidation was building within the organization. "Yes, everyone was concerned," School recalled, "but it wasn't until 2000 when [a few of us] attended a local group meeting in Michigan that we decided to take some action." That meeting was held in late May at a golf outing at the Grand Haven, Michigan, Country Club hosted by the West Michigan Rubber Group. Mark Petras was assistant secretary of the Rubber Division and drove over to the meeting from South Barrington, Illinois, just northwest of Chicago.

"I know we had more than the usual number of attendees because it was a golf outing," Petras recollected. "It was a typical late spring/early summer day and we held our meetings at the Holiday Inn in Spring Lake."[22] Nothing out of the ordinary occurred at the meeting, but outside of the formal structure—in the forum of a social convocation—the winds of change were gathering momentum. "With the increasing amount of time that people had to devote to the job, it was becoming very difficult to find volunteers to serve as Rubber Division chairman," Petras said. "Employers were becoming more reluctant to allow their employees to spend time outside of their jobs. Since the division depended heavily on volunteers, this trend, and the other financial issues were impacting our operations."

After that meeting in Michigan, School revisited his 1988 note and reiterated the contents to the steering committee, which included Chairman Toms Royal, former chairman (1994) Tom Dendinger, Probasco, Nygaard, and Petras. The revelation also was made to all area directors of the Local Rubber Groups. "The Rubber Division was—and is still—a member-driven organization," recalled Royal. "For the first forty years, it operated primarily on dues from members who came from a growing

industry. The management of the Division was all volunteers and programs were driven by elected members of the executive committee. In
order for the Division to operate efficiently, the rubber industry companies
granted time to their employees in support of the Division. It helped us to
maintain a sharp edge in the technical developments of polymers [of the
rubber discipline]."[23]

Within all the legal constraints that a nonprofit professional organization must operate, the Rubber Division had developed in the mid-
1960s an event to help them supplement the income received from
dues and semiannual technical meetings. These were called "expos."
"The concept of the *Rubber Expo* and the *Mini-Expos* was to allow the
Division a source of income to expand programs," said Royal. "Volunteers ran the technical meetings from the area where the meetings were
held, but more services were needed, and employers were less likely to
allow the volunteers enough time to run all of the burgeoning Division
activities." The Rubber Division's expos, or trade shows, were born in
the spring of 1965, at the Fontainebleau Hotel in hot and steamy
Miami, Florida.

Ralph Graff is one of the true veterans—some say the consummate salesman of and for the Rubber Division. Jim D'Ianni, who was only one of
three people who have held the position of both president of the American Chemical Society and chairman of the Rubber Division,[24] said that
Graff, "is a wonderful person. If anybody merits more recognition [for
his work for the Rubber Division], I think that Ralph is right up there."
Graff was born in Johnstown, Pennsylvania, on New Year's Eve 1923 and
realized from his youth that the local steel mills weren't in the direction
of his career path. After a short stint in the U.S. Army during World War
II, he returned to Pennsylvania and the University of Pittsburgh, where
he completed his four-year degree.

Graff first became familiar with the rubber industry in 1946, when he
started working for Goodyear in Akron. A year later, he joined the
Akron Rubber Group, one of the first local rubber groups formed in
1928. An opportunist extraordinaire, Graff "went to the Rubber Division
meeting in 1949 because some of the carbon black people sponsored
some Goodyear people up there," he said, laughing.[25] Graff joined the

American Chemical Society in the same year, and in 1950, he joined the Rubber Division.

In late 1949, Graff left Goodyear and "moved to Wilmington, Delaware first, and then sought out DuPont and was able to sell myself to them," he boasted. At DuPont he shared an office with Seward Byam, who was active in the Philadelphia Rubber Group. In 1953, Byam became chairman of the Rubber Division and asked his good friend Graff to help out on the local arrangements committee when the division met in Philadelphia. It wasn't long until Graff became chairman of the Philadelphia Rubber Group in 1959, and he claims that during a period from 1955 to 1992, he never missed a Rubber Division meeting. "Well, maybe I missed one in Providence a few years ago, it was the first one I'd missed in fifty years I think," he said. "My bosses always would ask me, 'Are you going to spend any time this week with DuPont?'"

Graff has the unique distinction of being the last individual in the Rubber Division to achieve a twelve-year succession of service—beginning in 1967 as an advertising manager of *Rubber, Chemistry and Technology* for three years, three years as assistant secretary and assistant treasurer, and then three years as treasurer. He followed that post with chairman-elect, chairman, and past-chair. He also served as director of administration and then director-at-large, in addition to moderating several symposia and committee meetings. "In those days we were slow learners," he laughed again. He was on the steering committee from 1967 to 1992, and during that time frame, in 1980, he held the chairman's job.

Likely since the birth of the division, suppliers and vendors—for several good and legitimate business reasons—had been courting most members of the Rubber Division but were continuously looking to increase their presence, especially in forums where a great number of attendees gathered. Those opportunities were only available once a year, but that began to change in 1964. "When some of our largest manufacturers—especially the tire companies like Goodyear, Firestone, General, Uniroyal and B.F. Goodrich—heard we were meeting in Miami Beach [in 1965] they weren't particularly pleased," said Graff. "They wanted to display equipment, but the cost to transport their goods down there was pretty steep. It was Jim D'Ianni, in his prominent position at

Ruth Murray (*left*), who became the Rubber Division's librarian in 1970, and Marge Bauer were on hand for the 1976 dedication ceremonies of the John H. Gifford Memorial Library and Information Center. The library is now located in the Norman P. Auburn Science and Engineering Center. (Photo courtesy of *Rubber World*.)

Rubber expos were not only a business focal point for the Rubber Division but also a welcomed social activity for the members. In 1981, longtime division administrator Marge Bauer handles the tradition of cutting the ribbon to open up the exhibit floor at the Cleveland Convention Center.

At the 1981 Rubber Division spring meeting in Minneapolis, the entire executive committee (*front row*) and steering committee (*back rows*) gathered for a rare photo opportunity. The two female attendees in the front row are Marge Bauer (*left*) and Barbara Hodsdon of the ACS staff. (Photo courtesy of University of Akron Archives).

Don Smith is current editor of *Rubber World* magazine and has been with the publication for thirty years. In that time, Smith said he has attended every annual Rubber Expo/Mini Expo except one. This is a photo of Smith at the 1997 Rubber Mini Expo in Anaheim, California.

Drawing from his own personal tragedy, longtime Rubber Division member Charles Rader (*left*) planted the seed for a division-sponsored scholarship program. Next to Rader is Jay Cline of Ferris State University, the 2006 undergraduate award winner. (Photo courtesy of Shelby Washko.)

On October 21, 1997, at the Cleveland Rubber Expo, the Rubber Division announced a one million dollar gift to help fund a new academic annex for the University of Akron's Olson Research Center for polymer engineering. The annex, now the site of division offices, added needed classroom and lecture space for the university's rapidly expanding polymer engineering program and allowed new laboratory uses of existing space in the Olson building. *Left to right:* University of Akron provost Noel Leathers, Stan Mezynski, Donald Mackey, Toms Royal, Frank Kelley, J. Marshall Dean III, University of Akron president Marion Ruebel, and Dan Hertz.

A native of Barnsley, England, Kent Marsden was asked to handle executive and administrative duties of the Rubber Division in January 2000 on an interim basis. Marsden dutifully handled the assignment and the year-long transition of the division to its present status with a full-time staff. (Photo courtesy of Kent Marsden.)

On August 29, 1998, the ACS officially recognized the Synthetic Rubber Program
with one of its national historical chemical landmark designations. The plaque,
placed in front of the Goodyear Polymer Center on the University of Akron campus,
honors contributions from five companies, government agencies, academic
institutions, and industrial laboratories that participated in the program from 1939 to
1945. Many Rubber Division members belonged to those groups. Ed Miller (*left*),
current division executive director, and Gary Horning, 2009 division chairman, are
shown next to the plaque. (Photo courtesy of Shelby Washko.)

Before the offices were moved to the present location, the Rubber Division operated on the third floor of the Goodyear Polymer Center on the University of Akron campus from 1991 to 2001. (Photo courtesy of the University of Akron.)

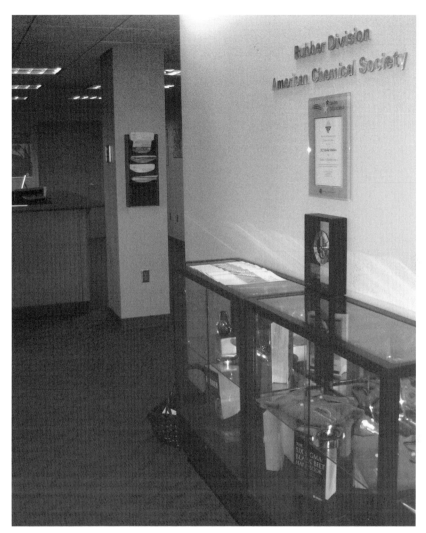

Offices of the Rubber Division on the fourth floor of the Polymer Engineering Academic Center on the University of Akron campus. (Photo courtesy of Chris Laursen.)

The Polymer Engineering Academic Center on the University of Akron campus is the present home to the offices of the Rubber Division. (Photo courtesy of the University of Akron.)

The 2009 Rubber Division staff. *Front row:* Christie Robinson, Christopher Laursen, Melanie Avdeyev, and Karen May; *back row:* Lakisha Miller, Missy Beynon, Sue Barr, Linda McClure, and Ed Miller.

Goodyear [director of research and development], who convinced them to go. Then the others followed suit."

D'Ianni and the division leaders also were wise enough to get tacit approval from the membership before making the decision to go to Miami. "But we didn't have ten thousand dollars to conduct a survey," Graff said, "so we made calls to several area directors where there were manufacturers to see what they thought. They liked the idea of a dedicated expo for the key suppliers, vendors and manufacturers, so we made the decision to go to Miami Beach in the spring of 1965."

The key companies attending were prominent and represented some of the best-known organizations in the industry. Names such as DuPont, Naugatuck, 3M, Cabot, Farrell, Desmond, Schwinn, and Shell were prominent in Miami, and at several other exhibits that followed the Florida event. However, even with the okay of the membership, Graff knew the division had to go one extra step to assure them of a successful show in the Sunshine State. "We were concerned that a lot of people would be spending time at the beach," he said. "Well, there was a canopy area right next to the walkway that led to the beach. We decided to put the exhibits under that canopy so everyone who went to the beach had to go by the exhibits!"

Graff said that the Miami meeting also was the launching site for one of Shell's new products. "They had a booth to introduce Kraton, a thermoplastic elastomer," he said. "They made rubber bands out of it, but TPE didn't make very good rubber bands in the hot sun. They were stickin' to everything, but the product turned out to be great." Today, Kraton is a highly successful product that offers many of the properties of natural rubber, such as flexibility, high traction, and sealing abilities, but with increased resistance to heat, weathering, and chemicals. It is also used to fill spaces in optical fiber cables to inhibit penetration of water or water vapor. The name has become synonymous with the cables in which the compounds are used.

Ten years later, on yet another crafty venture, Graff teamed with fellow Rubber Division colleague, Tom Loser, founder of Wyrough and Loser, and sold dispersions out of their offices in Trenton, New Jersey. Graff was division treasurer at the time and, as such, was constantly cognizant of keeping the ink on the ledgers always black. "We were having such a great deal of success reprinting [technical] papers and selling copies at the Wire Cable meetings," Graff said, "that we started doing [it] at Rubber Division meetings, too."

However, Graff, the noble efforts of others to focus on the bottom line, and the expos still didn't provide the *deus ex machina* that was required to steer the vessel in the right direction. Reinforcing Rudy School's 1988 vision, Toms Royal said, "In order to grow, we needed a structure in place—outside of the state government—where we could control our own destiny."

―――――――――

Barnsley is in the north-central part of England, about sixty miles east of the famous coastal town of Liverpool and about eleven miles north of Sheffield. Chartered in 1249, Barnsley, referred to as the "Tarn" by many of its residents, was built on coal and people of strong backs and acute minds. The borough's Latin motto is *Spectemur Agendo*, "Judge Us by Our Actions." Charles Marsden, who made his living as a lathe operator and as a coal miner in South Yorkshire, likely was a man wise beyond his peers. He knew that there was no future for him and his family in Barnsley, especially in coal mining. That's why, in October 1955, he and his wife Sarah Dickson Waddell, their eleven-year-old daughter Kay, and nine-year-old son Robert Kent sold all of their possessions, boarded the *Queen Elizabeth I*, and sailed to America.

Charles's wife Sarah had an uncle and aunt, Bill and Mary Wright, who lived in the United States and sponsored the Marsdens. Several years prior, the Wrights themselves had gravitated to a place called Akron, Ohio. Charles became comfortable with Akron almost immediately. This handyman and former miner found his first job with Burger Iron on South Main Street, but within a few months he took his skills to the Goodyear Tire & Rubber Company on East Market Street—Department 111c, Machine Design.

Charles's young son Kent loved his new home almost as soon as he boarded the *QEI* in Liverpool. "I remember the trip fondly," he recalled. "I had the run of the ship. I had no regrets about leaving the country and just wondered why everyone from Barnsley to the Southampton boarding docks was crying."[26] He didn't know it at the time, but almost forty-five years later, Robert "Kent" Marsden would play a vital role in the development of the present structure of the Rubber Division of the American Chemical Society.

Barnsley, England, where Kent Marsden spent the first nine years of his life, is some three thousand five hundred miles away from him and many memories ago. Kent's sister Kay Lossing lives in North Canton. His father retired from Goodyear after thirty-one years at Plant 1 but died in 1997. His mother Sarah died in 2005. Goldthorpe, the last coal mine in Barnsley, England, closed in 1994.

Robert Kent Marsden was named after his uncle, his mother's brother Robert, a British soldier, and Kent, a bunk-buddy of Robert's in a German concentration camp during World War II. "Uncle Robert would always send us post cards and sign them 'Robert and Kent,'" said Marsden. The many postcards from Robert became mental pictures so when Sarah had her second child in Sheffield, she named him Robert Kent. Robert survived the concentration camp, but his friend did not.

Kent Marsden loved America, Ohio, and his new home. He grew up in Ellet, where he quickly banished his British brogue and assimilated into American culture with relative ease. About the only thing Kent retained from England was his skill and love of soccer. He attended Hyre Junior High School and graduated from Ellet High School in 1965 and the University of Akron in 1970 with a B.A. in secondary education and three varsity letters in soccer. In 1971, he married his high school best friend and college sweetheart Pamela Kay Slee. After a short stint in the Army with the 82nd Airborne Division during the Vietnam War, and a whistle stop in the Akron public school system, he returned to the University of Akron in 1984, this time as an employee, director of corporate and foundation relations, in the Development Department.

In 1996, he became director of administrative services for the College of Polymer Science and Polymer Engineering, working for the dean, Frank Kelley. In late 1999, his job description expanded to include some unwritten assignments. "Frank came to me and said, 'Kent, could you do me a favor? Can I assign you to another responsibility for a period of time?'" Marsden recalled. "And I said, 'Of course I'll do it.' I do things."

Kelley, who had financial and managerial responsibility for the office personnel of the Rubber Division, asked Marsden to temporarily manage that office—hands-on—while the office manager was on medical leave. After the holiday season in 1999, Marsden officially became the "landlord" of the Rubber Division of the American Chemical Society on January 2, 2000. There was no twenty-one gun salute or even a press release, and he didn't get much of an opportunity to acclimate himself to his new area of responsibility, nor did he receive executive training for his new position. "I had literally two hours with Lu Ann Blazeff to know what was going on over there [ironically, across an entryway from his own office], to know what I needed to do, then she left," said Marsden. "She did a couple of things at home for me that I had no clue how to do. One of them, for example, was the report to the American Chemical Society—the annual report of the Rubber Division. So I took on that responsibility, and, you know, I still had a full-time job—assistant to the dean."

For at least the next year, Marsden was asked to stay on by the division to assist in the day-to-day supervision of the Rubber Division and handle the transition, which included management of the personnel, finances, and so on, and coordination of meetings and the expos. "Toms Royal [the new chairman in 2000] and I had a huge challenge—to keep the business going," Marsden said. "Because a couple of months later, we had a meeting in Dallas, and I had to help put the Dallas meeting together with people that I didn't even know what their job was, what was their job responsibility. So I had a real huge learning curve to get up to speed real quick and pull off the Dallas meeting, and there were multiple things on my plate at the time . . . that I just had to learn and make sure that Toms Royal had a successful chairmanship of the Rubber Division."

The dedication of Royal to Marsden was reciprocal. "Kent managed the office needs and the meetings as the Division struggled to come up with a plan for . . . going forward," Royal said. "Based on the problems that existed, the executive committee and the officers were open to ideas on how to proceed. We also hired an attorney, Karen Butera, to follow our footsteps."

The spring meeting in Dallas was conducted as smoothly as one could expect. Spring meetings are typically smaller than the fall ones, and while some technical papers are presented, these meetings are needed for division governance to properly function. Still, the issue of a

new direction in leadership would continue to escalate even after the
meeting in Texas. "I found the ninety-nine dollar round trip airfare from
Baltimore to Cleveland and was at the Rubber Division office every two
weeks," Royal said. "Gradually things took shape as a plan, but it had to
be approved by the Executive Committee and implemented."

The dilemma did not deal with whether the division should change
or not, but how change could be accomplished smoothly, within legal
guidelines, yet with compassion. "We had a core of nine full-time, dedicated employees with jobs that often did not fit the new job description
and experience we were developing, nor did they fit the mold of the
available university jobs outside of the division," Royal said. One of the
first critical steps, however, of the planned new organization was to find
a leader.

———————

It was an untypical snowy, freezing evening on Saturday, December 2,
2000, in Lexington, Kentucky, but it was going to be a warm evening by
several measures for Ed Miller and his wife, Diane. The weekend provided the first good opportunity for the couple to celebrate a few recent
memorable events in their lives. Just two weeks prior to that, on November 12, their adopted, six-year-old, special-needs daughter Elizabeth had
arrived. The date, coincidentally, was also their son Joey's eighth birth-
day. Furthermore, the Miller's thirteenth wedding anniversary was on
November 14, and the occasions lent themselves quite appropriately to
a celebration.

Ed also considered November 14, 2000, a memorable date for another
reason. That evening he had just completed surfing the Internet, stop-
ping at one of his favorite sites that listed executive job openings at nonprofit organizations. One opening intrigued him. It was almost an exact
match of his present position as president and CEO of the Asphalt Institute. So he attached his updated resume and replied to the ad. A week
later, he received a reply, but the company placing the ad wanted him
to resend the material.

In the meantime, Ed and Diane wanted to celebrate. "We went dining and dancing with some friends," Miller recalled. "We were having a
good time and, of course, it was noisy, but by 10:30 we had had enough
so we decided to call it an evening." On the way home, Ed noticed that

someone had been trying to call his cell phone, but he didn't recognize the number. "Then the phone rang again while we were in the car, and I'll never forget the words that came from the caller: 'This is the fire department. We don't want to alarm you, but there's been a minor fire at your home. The kids are okay. How fast can you get home?' "[27]

The Millers rushed home, and when they pulled into their neighborhood, "lights were flashing everywhere," Miller said. "It was freezing, snowing. There must have been a dozen fire trucks, ambulances and every neighbor outside our house. It was like a big circus." Ed and Diane negotiated their way through the commotion, running to find their children. "We ran into the fire chief and he asked us, 'Are you the Millers?' We said, 'Yes, where's our kids?' "

The children, Joey and Elizabeth, were safe inside a rescue vehicle, albeit in pajamas and no shoes—but safe. The babysitter had called her father when Joey said he smelled smoke and then called the fire department. The entire basement was gutted, the first floor was burned badly, and smoke and soot permeated the house—even down to clothing in upstairs drawers. Miller said the fire gave him a more realistic outlook on life and his future. "One thing I really learned is that you had to let go real quick," he said. "It's only stuff . . . a lot of stuff, but it's only things. By letting go, it helped us to focus on what really matters. . . . We were blessed . . . that the kids were alive." The experience also helped him develop a perspective of what would occur next. Within a few weeks, he would become the first executive director and first full-time employee reporting directly to the executive committee of the Rubber Division of the American Chemical Society.

On January 15, 2001, slightly more than a year after Kent Marsden was given responsibility of the staff of the organization, the executive committee of the Rubber Division selected Edward L. Miller, who was president and chief executive officer of the Asphalt Institute in Lexington, Kentucky, as its first executive director, signaling the change of a new era in the history of the organization.

NURTURING HISTORY

15

A morsel of genuine history is a thing so rare as to be always valuable.

THOMAS JEFFERSON

Audio recordings have a pedigree dating back to the nineteenth century, when Danish inventor Valdemar Poulsen developed and patented a wire recorder called the telégraphone. However, it wasn't until the 1940s that the telégraphone's next generation, the magnetic tape recorder, was perfected by two German firms, AEG and a company familiar to the rubber industry, the immense chemical group of I. G. Farben.[1] Between 1940 and 1973, the tape recorder grew so much in popularity that six presidents of the United States, from both parties, took advantage of the device and secretly recorded more than five thousand hours of conversations.[2] On a smaller, less recognized scale, since 1964, three individuals from the Rubber Division of the American Chemical Society used the tape recorder to chronicle interviews of almost a hundred people in 169 hours of history-filled conversations.

Those modern day scribes are the Rubber Division's historians. Herbert A. Endres, Benjamin Kastein Jr., and Shelby J. Washko might not be as famous as the United States' chiefs of state, but what they documented on audio recordings has been intuitive and extremely provident, especially for the entire rubber industry. Over the years, their combined efforts have given the world an evergreen record of the storied one-hundred-year history of not only many prominent individuals in the Rubber Division but also the rubber industry as a whole. These three individuals have been the only official "oral" historians of the division and have recorded interviews starting in 1964 that border on being clairvoyant.

Their recordings document the verbal thoughts of Rubber Division participants and observers, some of whom, such as Arnold H. Smith, Warren K. Lewis, William B. Weigand, and Walter W. Evans, trace their participation back to the 1920s and were members of the original India Rubber Section and the founding fathers of the Rubber Division. There were even several individuals they interviewed who could recall their jobs, contributions, and so on prior to World War I. Other names of interviewees look like a "Who's Who" list of rubber chemistry pioneers and legends, including John M. Bierer, Ernest R. Bridgewater, Howard I. Cramer, Bradley Dewey, Jim D'Ianni, Raymond P. Dinsmore, Benjamin S. Garvey Jr., and Harold Gray. Edwin J. Thomas, Raymond Firestone, John L. Collyer, and M. G. "Jerry" O'Neil, magnates of four of the tire industry's largest companies—Goodyear, Firestone, B. F. Goodrich, and General—also were subjects of Rubber Division historians' interviews.

Three of the aforementioned—D'Ianni, Dinsmore, and Garvey— were also Charles Goodyear Medal winners, as were more than half of the recorded subjects. The historians' list of interviewees also includes eleven who were elected to the International Rubber Science Hall of Fame. Those interviewed by Rubber Division historians on audiotape were Melvin Mooney (selected to the hall in 1971), G. Stafford Whitby (1972), William B. Weigand (1984), Herman F. Mark (1993), Maurice Morton (1994), Norman Bekkedahl (1995), Adolf Schallamach (1998),

Robert McKee Thomas (1999), William J. Sparks (1999), Leonard Mullins (2001), and John D. Ferry (2003).

The motivation behind the idea to record history actually came from a "suggestion" by the American Chemical Society. "One year the Division of History [of the ACS] at that time made the recommendation to all the divisions to consider taped interviews to document historical events [in the society]," said Kastein. "I think one of our councilors brought that idea back from one of the meetings, and I think the Rubber Division was the only one that acted on it."[3] Fran O'Connor, who in 1965 had just moved to Akron to work for Harwick Chemical, believes Jim D'Ianni, the division's ACS councilor and chairman in 1964, provided the necessary motivation. "It's something that Jim would consider important," O'Connor said.[4] And the timing fit.

O'Connor was right. In a Steering Committee meeting in Washington, D.C., on November 14, 1963, D'Ianni proposed that Herb Endres and Sid Caldwell be appointed division historians. On February 10, 1964, the committee recommended that Endres alone be appointed, and on April 28, 1964, at the division's meeting in Detroit's Sheraton-Cadillac Hotel, D'Ianni made the proposal and recommendations official. However, division meeting minutes indicate that in 1955, Ed V. Osberg technically was appointed the first historian by Chairman John M. Ball on November 2 at their meeting in Philadelphia's Bellevue Stratford Hotel. However, a little more than six months later, Osberg, in a letter to the new chairman, Arthur E. Juve, resigned with no explanation.

The position of historian was vacant until May 14, 1957, when Chairman Benjamin S. Garvey asked B. F. Goodrich's Harlan L. Trumbull, who had served as division chairman in 1937, if he would take the job. Trumbull agreed, but four months later he too resigned. In September 1958, a year after Trumbull vacated the position, Firestone's R. F. Dunbrook reported that "no historian could be found to succeed Trumbull." The position was dormant—not even a single mention in meeting minutes—for five years until D'Ianni stepped to the plate.

Regardless of the progression, the division finally had a solid commitment to its past, and an official historian and "oral" histories became a reality from the brain trust of the organization, the Executive Committee. They selected a recently retired Goodyear chemist, Endres, manager of research applications, as their first official "oral" historian, of course with other customary documentary duties. Endres, an Akron native, received his undergraduate degree from Stanford University and returned home to Akron and the university, where he became one of the first students to receive a master's degree in chemistry. Endres, who was very thorough in his interviewing techniques, even had the foresight of recording himself on April 24, 1967, to explain the reasoning behind the endeavor:

> The archives program was conceived about 1963 by some members of the Division of Rubber Chemistry who were concerned that much valuable historical information about the development of rubber science and technology was not being made a matter of record. Many of the important contributors were still living, but were close to or past retirement age. The stories of the early pioneers—with only a few exceptions—were in danger of being lost forever. As a consequence, the archives program was set up to record the personal recollections and reflections of the part these people played in various important developments.

Endres approached his task with solemn professionalism and, in 1965, took off like a one-man radio news department with his Roberts reel-to-reel tape recorder—a genuine antique by today's standards—and interviewed seventeen people. Kastein much later explained Endres's urgency to conduct so many interviews in the first year by candidly saying, "We were losing a lot of distinguished scientists, and we wanted to record them before they passed on." The original criteria for the interviews, established by the division and the historians, remain fairly intact. Priority is given to Charles Goodyear Medalists, past chairmen or secretaries of the division, organizers of the local rubber groups, substantial contributors to the rubber literature, and fifty-year members of the American Chemical Society.[5]

Ironically, even before Endres could thread the tape on his recorder, the ACS's chairman on the Committee of Archives of the Division of the History of Chemistry submitted a tape-recorded interview to him to be used as a sample or template for the division's own efforts. On October 9, 1964, Wyndham D. Miles, who also held the title of historian of the National Institutes of Health, interviewed John C. Bailar, president of the

ACS in 1959, who had just won the 1964 Priestley Medal, the highest award conferred by the ACS. Miles also followed that interview with a recording on January 6, 1965, of B. F. Goodrich's Harlan Trumbull, a fifty-year member of the ACS and a charter member of the Akron section of the society. Miles conducted his last interview for division archives on February 2, 1965, with Per K. Frolich, president of the ACS in 1943.

Endres began the official division-sanctioned interview process that exists today by taping an introduction of Miles's interview of Trumbull on January 16, 1965. His initial one-on-one interview was conducted on February 8, 1965, with Raymond P. Dinsmore, who was Rubber Division chairman in 1927 and winner of the 1955 Charles Goodyear Medal. Dinsmore was vice president of research at the Goodyear Tire & Rubber Company and served on the board from 1960 to 1964. As was Endres's signature, he usually added a personal comment at the end of most of his interviews. For Dinsmore's case, he said that close friends of the Charles Goodyear Medalist referred to him as "Dinny."

Endres was born in Akron in 1897, and with the exception of his time in California earning his B.A. in chemistry at Stanford and working for a short period at the California Institute of Technology, he spent most of his life in the Rubber City. He returned to the University of Akron, where he received his master's in chemistry, and took a job at the Goodyear Tire & Rubber Company in research applications and services. He retired in 1963 after thirty-one years.

"He had a fairly good sense of humor, but, ironically, he wasn't very talkative," said Leland "Lee" Endres, Herbert's only son and a professor of chemistry at California Polytechnic State University in San Luis Obispo, California. "He would particularly like to smoke a cigar in meetings and hold it in his hand until the ash was one-to-two inches long just to make everyone nervous." Lee said his father was very active in the Boy Scouts, Silver Lake Church (where he said he was a participant in its founding), and the Village of Silver Lake (where he joined the fight to improve the water system and build a water tower) and "was quite proud of his lawn. I wasn't that close to my father, but a while back I was conducting a search on one of my papers in Chem Abstracts, and there, right on the same page as mine, was one of my father's papers," Lee said proudly. "I bet you don't see that very often."[6]

Endres held twenty-one U.S. patents. He died on September 18, 1977.

The recognized sage of the historians, Benjamin Kastein Jr. retired as manager of dry rubber technical service from Firestone in 1978. That same year Rubber Division chairman Frank B. Floren of Exxon Chemical asked him to take over from Endres. Kastein was uncharacteristically "kind of reluctant about doing oral histories," he said. "But once I got into it, it was a very fascinating part of the duties of the historian." He was so fascinated that he held the position for fourteen years, one more than his predecessor, Endres. Shelby Washko, who retired from Polymerics as director of sales in 2007, was handed the cassette recorder in 1992 as the division's incumbent historian, and she continues to add years to her record as the resident historian with the longest tenure.

The title "historian of the division" was loosely interpreted as the organization's "great communicator," and Kastein was a perfect fit. Ben had a different interviewing technique than did his predecessor. Endres always used a set of questions he prepared in advance of his interviews, and often included a prewritten text that either preceded his interviews or concluded them. Yet Kastein was the precursor of extemporaneous or "off-the-cuff" interviews for the division's taped archives.

Kastein was born in Waupun, Wisconsin, on August 7, 1916, and received his B.S. in chemistry from the University of Wisconsin in 1938. His first job after he graduated was as a chemist at Shaler Company in Waupun, but in 1940 he was offered a raise of twenty-five dollars a week—to $125—to move to Akron, Ohio, and join the Firestone Tire and Rubber Company. He became a member of the Akron Rubber Group and the Rubber Division in 1941, has held several positions within both organizations, and was division chairman in 1975. For many years Kastein was a technical editor of *Rubber World* magazine and authored many technical papers, but the one that defines Kastein the best is not technical at all. It's titled "People Make the Difference," and while it is a concise history of the Rubber Division, its worth is in revealing Kastein's insightfulness.

His philosophy is his banner. "The division is all about relationships and information," he said. "We get probably more information in the hallways than in the sessions, and we'd share that information with our counterparts at the other rubber companies." They weren't doing it for covert activities because they knew that every compounder that was worth a gram had his own "black book" in his back pocket anyhow, and he guarded that with his life. It was just the nature of the scientist or researcher to swap stories or techniques. After all, the division was formed in 1909 because twenty-eight men wanted to share information, and "it's been true for a hundred years," Kastein said. He knew that, and lived by it, so it fit his personality and technique perfectly to interview colleagues for posterity.

It was fortunate for Kastein that Frank Herzegh, the Charles Goodyear Medalist in 1978, was his first interview. "It was the most outstanding of the whole series I did," he admitted. "He was just an interesting individual." Herzegh, who worked for B. F. Goodrich, received the medal for his work in perfecting the tubeless tire. "They reached the point where they thought they had the correct control tires and wanted to test them in Texas," Kastein said. "And Herzegh wanted to drive the car. On the test run, one of the control tires had a blowout and Frank lost control of the car. It spun out into a ditch, but in those days they didn't have seat belts, so Frank held on to the front seat. He walked away from the demolished car and went back and reported to his superiors, 'the tubeless tire was a tremendous success, but the car wasn't.' Then he handed his superior all that was left of the car, one hub cap."

When he handed the microphone to Washko in 1991 after conducting his last recorded dialogue with Charles Goodyear Medal winner Edwin J. Vandenberg, chairman of the Division of Polymer Chemistry of the ACS in 1979, Kastein had interviewed forty-one people whose lives and achievements had made an impact on the Rubber Division and its members. Some of those interviewees were former United Rubber Workers (URW) union health and safety consultant Louis Beliczky; URW president Peter Bommarito; Mel Lerner (former editor of *Rubber Age* magazine); Ernie Zielasko (former editor and founder of *Rubber & Plastics News* magazine); and former company chairmen Ray Firestone (Firestone), M. G. O'Neil (General), and E. J. Thomas (Goodyear).

While the librarians were the solid research bases of the division, if any member had a question about precedent or needed a morsel of anecdotal information to impress a supplier or vendor, the historian was the individual to corner at a cocktail gathering. The historian also inherited the unofficial title of division photographer. Fran O'Connor, former division chair in 1974, preceding Kastein, said that Ben took his job as photographer very seriously. "He even took photos at my daughters' weddings," O'Connor said.

Shelby Washko also enjoys the photographer's hat that she now wears as part of the historian's job. Before she retired in 2007 from Polymerics in Cuyahoga Falls, Ohio, as director of sales (her second retirement, having retired from Zeon Chemicals in 2000), she was also the company photographer. However, to keep her skills really honed, for "practice" every year during the Christmas holidays, she uses the inside of her home as a model for many photos she takes documenting the hundreds of decorations she prepares for family, friends, and herself. "I truly enjoy this every year," she said. "I decorate differently every year and it's a challenge . . . but I start around Thanksgiving and I love every minute of it."[7]

While Endres and Kastein set a solid precedent on interviewing skills, Washko took the historian's role to yet another level. Washko, born in Deepwater, West Virginia, took a different path than most of her "fellow" colleagues to her position in the division and in the industry. She began her career in 1962 as a buyer in the Purchasing Department, later moving to an administrative position at Goodrich-Gulf. "[The company] was the result of a merger between B. F. Goodrich and Gulf Oil," she said. In the late 1960s, B. F. Goodrich bought out Gulf's interest and became B. F. Goodrich Chemical. Shelby moved to an administrative position at Goodrich Chemical. "I was working during the day, going to college at night, had kids and a full plate and had just interviewed with SOHIO to go into purchasing for them," she recalled. "In 1976, when I told my boss I was planning to leave, they offered me a job in sales, but I said 'I don't have a technical background.' They said, 'that's okay. We'll teach you.'"

Shelby went out to Goodrich's Avon Lake, Ohio, facility and literally got her hands dirty learning how to compound rubber and cut it on a mill. She mastered the mill on her second try, an implausible path to sales. Yet months later she was handed a territory worth two million dollars, and "the first thing that happened to me at my new job," she said, "Tony Luxedor, then membership chairman for the Rubber Division, came to my desk and gave me an application and set it down and said,

'Here. You need to join the Rubber Division.'" She heeded his advice and made a giant step forward.

The first interview she conducted as division historian in 1992 was with Charles Goodyear Medal winner Ronald S. Rivlin, who conducted pioneering studies of the elastic properties of rubber and developed formulas to explain their behaviors. Rivlin was born in London and earned his doctorate from Cambridge University in 1952. He eventually became professor emeritus in applied mathematics and mechanics at Lehigh University. "I was very impressed with his credentials," said Washko. "I enjoyed the interview, but I became more comfortable as everything progressed." While the conversation with Rivlin was notable, Washko had her favorites. Among those was the interview she conducted with 1997 Charles Goodyear Medal winner Adel Halasa. "I just loved the interview with Adel," she said. "He had a very humble background and it was extremely interesting. He was very jovial and talked a lot about his grandmother."

Halasa, a research chemist from Goodyear, was born on December 24, 1933, in Madaba, Jordan, one of twelve children, and grew up "seven kilometers from Mount Nebo . . . drinking the water that flowed down the mountain," he told Washko. "My mother died giving birth to my twin sisters. She hemorrhaged badly and we didn't have doctors at the time. I was actually raised by my grandmother and father, who chose me to be educated." Halasa came to the United States when he was seventeen years old and migrated to the University of Oklahoma in Norman, where he met and married his wife Ofelia, and graduated in 1955. He earned his doctorate in chemistry from Purdue University in West Lafayette, Indiana, and took a job at Firestone in 1964. In an uncommon tire industry occurrence at the time, he left Firestone in 1979 and accepted a five-year government-contract position in Kuwait.

However, when Halasa wanted to return to the United States in 1984 ("I didn't realize how much of an American I really was"), Firestone offered him his job back—but added that they were reducing their staffs. Halasa did not want to be a victim of downsizing, so he called an old friend, Carl Snyder, head of research at rival Goodyear, who offered him "more freedom to operate." Adel took Snyder's offer and became the lead researcher on the world's first aqua channel tire, the Aquatred (introduced in 1991), developing a new compound, styrene isoprene butadiene rubber, or SIBR.

Other Washko favorite interviewees included Aubert Coran, the Charles Goodyear Medal winner in 1995. "I took photos of him and his family at the Philadelphia Mint. The mint was beautiful and provided outstanding marble backgrounds for professional-looking photos," she said. "Jack L. Koenig, 2000 Charles Goodyear Medalist, a former research chemist at DuPont and the Donnell Institute professor emeritus at Case Western Reserve in Cleveland; and Leo Mandelkern, 1993 recipient, and a leader in polymer education were extremely enjoyable interviews," she added. The interview with Mandelkern is significant as it was the historians' first attempt to use video. "Dr. Mandelkern was a very humble man," Washko said, "and was very cooperative as I tried to videotape him, too." However, the video experiment ended as soon as it started. "We ran out of tape," she said with a laugh. "Besides, it got in the way of me taking pictures."

Washko added another touch to her function as historian. "I would follow them around all week and take photos of them and their spouses at the various activities and meetings during the conference and would give them a photo album after it was all completed," she said. The album project, while well accepted, met its demise when digital photography was introduced in 2004. "But that can be a problem, too," she admits. "One time, photos in the camera were accidentally deleted." So much for modern technology.

While the division depended on the historians for the majority of the interviews, on a couple of rare occasions they had some ad hoc assistance. In addition to Wyndham Miles's two prototype interviews, in 1976, Earl Gregg, division chairman, and Jim D'Ianni were interviewed by a KTSF radio reporter in San Francisco for the meeting the division held there that year. A DuPont public relations representative recorded the session with Arnold Collins in 1981, and Roger Richardson of Esso, also dean of the engineering college at Louisiana State University in Baton Rogue, interviewed his former colleague Robert McKee Thomas.

The contributions of others notwithstanding, Endres, Kastein, and Washko are a manifestation of the division's long-term commitment to the preservation of events and individuals who have had an impact in the history of rubber, rubber chemistry, rubber research, and science in general. In the self-recording he made to explain the reasoning behind the endeavor, Herb Endres offered a fitting conclusion: "By continuing this program from year to year, a very interesting history of rubber science and technology in the words and voices of those who created it will be made available for posterity."

BORN
OF TRAGEDY

⟨16⟩

*It is, in fact, nothing short of a miracle that the
modern methods of instruction have not
entirely strangled the holy curiosity of inquiry.*

ALBERT EINSTEIN

Charles Andrew "Andy" Rader started swimming when he
was three years old. When he graduated from Copley High
School in Ohio in 1983, he had earned four varsity letters
in swimming and a district championship, and during the
summer, he served as a swim coach at Fairlawn Country
Club. Most important, Andy had been accepted to attend
college at Indiana University in Bloomington. While he
didn't receive an athletic scholarship, he made the swim
team at Indiana as a walk-on and swam on the varsity squad

until he graduated with his unique double major in marketing and biological science.

Andy loved swimming, but he loved his college sweetheart, Susan Gay of Chattanooga, Tennessee, even more. In 1992, shortly after he graduated, Andy and Susan were married, and he was hired by Eli Lily's Dista Products. Susan took a job at Saint Vincent Hospital in Indianapolis as a nurse practitioner. Andy had accomplished a great deal in his short time with Dista and was assigned to a team that successfully marketed and sold the nationally known antidepressant drug Prozac. His group set several area sales records. By many measures, the couple was successful and very happy. The world was their proverbial oyster.

On the night of December 8, 1995, Andy and Susan were returning by car to their Pendleton, Indiana, home, which they recently had purchased, and didn't notice that the roads had become treacherously iced over. When their vehicle hit the "black ice," it spun out of control. Both Andy and Susan were thirty-one years old when they died in the tragic accident. Andy was the son of Clarita and Charles Rader of Akron, Ohio. Charles was chairman of the Rubber Division in 1986 and, at the time, a member of the division's executive committee and a councilor to the American Chemical Society. "The Indiana Highway Patrol called us the next morning and I didn't know if it was real or a prank call," said Charles Rader. "I said 'give me your number and I'll call you back.' I did. I asked Clarita to pinch me to see if it was real."[1]

Even for the former University of Tennessee standout college football tackle who was a three-year letterman and an academic all-American in 1956, Charles Rader admits that it was extremely difficult for him to accept the loss of his son and daughter-in-law. "My wife and I were in the depths of despair," he said. A few weeks later, though, the Raders received a call from two of Andy's former high school swim team members. Bob Simmons and Bob Beiswenger wanted to have the family's permission to establish a college scholarship in Andy and Susan's memory to be given to a Copley High School graduate.

"We were touched and grateful," said Rader. "So together we contacted the high school to see what we had to do." They made a fortuitous call to the guidance office and talked to Keith Walker, the counselor. "The Raders took the high road," said Walker, now retired after spending thirty-six years in the Copley-Fairlawn School District teaching English and physical education. "The Raders wanted to give back something to the community in Andy and Susie's memory. We were more than pleased to help them."[2]

Walker and his colleagues helped the Raders and their daughter Marie establish the Andy and Susie Rader Memorial Scholarship through the nonprofit Copley-Fairlawn Schools Foundation. With initial donations from the family and friends, the Raders set up a $20,000 endowment fund to offer a $1,000 annual scholarship. In the spring of 1996, the first recipient was Sonya Kapusinski. Since 1992, the Copley-Fairlawn Schools Foundation has awarded more than eighty scholarships totaling over $119,000. The compassion and thoughtfulness extended by the Copley students who made the initial request, the guidance counselor, and the Copley-Fairlawn High School community also served as a catalyst for another benevolent idea. Rader thought of other scholarship opportunities for high school students, contemplating how his own professional organization could play a contributory role. Then he approached the Rubber Division with this initiative. "I thought that we, as a division, should consider establishing a scholarship program of our own," Rader said. "So I bounced the idea off the Steering Committee and they were fully supportive of it. They suggested that we get some technical guidance in developing such a program."

So Charles returned to the people who had given him the idea in the first place—Keith Walker and his team at Copley-Fairlawn High School. "He and the staff were very cooperative and helpful with the professional guidelines for such an effort," Charles said. "We developed a program for the Rubber Division and sent the proposal to our treasurer at the time, Toms Royal, and he liked the idea. We received tentative approval, approached the Executive Committee, addressed a few of their minor concerns and modified it." The criteria for the scholarships were fairly simple, and not surprisingly, similar in structure to the guidelines established for the Andy and Susie Rader Memorial Scholarship at Copley High School: "Applicants may have a major area of study in chemistry, physics, chemical engineering, mechanical engineering, polymer science or any other technical discipline of relevance to the rubber industry. The scholarship winners will have a serious interest in full-time professional employment in the rubber industry upon graduation and shall show promise of a productive future as a technically trained professional."

The efforts of Rader and several others within and outside of the organization culminated in the Rubber Division's first scholarship program. In the spring of 1998, two years after the Andy and Susie Rader Memorial Scholarship was initiated, the Rubber Division awarded three undergraduate scholarships: to Stacey R. Magill at Pennsylvania State University, Holly C. Russel at the University of Cincinnati, and Joy A. Untch at the University of Akron.

Before Ed Miller became executive director, on October 19, 2000, at a meeting of the executive committee in Cincinnati, Ohio, the division approved provision of $100,000 of investment money and considered future donations to endow the Ferry and Flory Fellowships until they were fully funded. In April 2001, the finance and budget committee approved the funding.

The John D. Ferry and Paul J. Flory Fellowships, established in 1988, were the division's first involvement in offering financial assistance to students. These fellowships, however, were for postgraduate studies and were awarded in offsetting years for forty thousand dollars each over a two-year period "as long as grades were maintained." Richard Vargo at the University of Akron was the first recipient of the Flory Fellowship in 1988–89, and David B. Adolf at the University of Wisconsin was the first recipient of the Ferry Fellowship in 1990–91. "In early 2003, Charles Rader was asked to evaluate both programs," Miller said. "He found that only one graduate fellowship winner had gone into the rubber industry, while more than 90 percent of undergraduate scholarship winners had gone into the rubber industry. Based on this, a motion was made to the steering committee at its July 22, 2003, meeting in Cleveland to terminate the graduate Fellowship Program. It was approved, and since then, the scholarship endowment has been known as the Scholarship Foundation and continues to grow for future undergraduate scholarships. In the interim, we continue to fund two or three undergraduate scholarships each year using the annual operating revenues."[3]

In 2003, the division renamed one of its three undergraduate scholarships the Dr. Emmanuel Kontos Scholarship in memory of the contributions the forty-four-year veteran made to the chemical and rubber industries. Kontos died on October 21, 2002, shortly after his retirement from Crompton/Uniroyal Chemical. Kontos was active in the Rubber Division and was a professor of polymer chemistry at Southern Connecticut State University in New Haven. The recipient of the first Kontos scholarship was Kathleen I. Frank, a student study-ing rubber technology at Ferris State University in Big Rapids, Michi-gan. Frank also was president of Ferris's student chapter of the Rubber Division.

A few colleges such as Ferris have opted to take more of an active role in the education of their students who are interested in rubber and related subjects. In addition to Ferris, the University of Massachusetts at Lowell has an operational student chapter of the Rubber Division. Fer-ris State is supported by the West Michigan Rubber Group and UMass-Lowell has the New England Rubber Group as a mentor/guide.

The University of Massachusetts-Lowell has been recognized by the rubber division since October 10, 2002. The group continues to grow and has increased its membership from twenty-two in 2005 to thirty-six in 2006. The city of Lowell also has some historical significance to the Rubber Division. One of the early chairmen of the division, Ray P. Dins-more, who held the position in 1927, was born just outside of Lowell in Tewksbury, Massachusetts, and graduated from Lowell High School. In 1914, after receiving his doctorate in chemistry at Massachusetts Institute of Technology, Dinsmore went to work at Goodyear in Akron for factory manager Paul W. Litchfield. His first job was as a "lowly designer of gadgets in the experimental engineering department," he recalled in an audio interview in 1965 with division historian Herbert A. Endres, ten years after receiving the 1955 Charles Goodyear Medal. Dinsmore was instrumental in forming the Akron Section of the American Chemical Society so that members from Summit County "would not have to go to Cleveland," which has its own ACS local section.

The Rubber Division supports these college groups by annually pre-senting an Outstanding Chapter Award in addition to grants from fifty to five hundred dollars for new projects and community involvement projects. "In addition," said Miller, "we are in the process of partnering with the ACS Division of Polymer Chemistry to jointly establish a new chapter at Clemson University in South Carolina. This is particularly significant because we continue to partner more with this division as the rubber industry evolves into other materials as well as rubber."

The division's oldest continuous educational assistance program is the Rubber Technology Award. Originally established in 1991 as the A. Wayne Place Memorial Award, the Rubber Technology Award is designed for technicians and production personnel in the rubber industry. It was originally created in memory of the former chairman of Jasper Rubber in Jasper, Indiana. This scholarship provides an annual stipend to employees of the rubber industry to further their education in polymer technology. The grant provides remuneration to rubber industry employees to allow them to attend a spring or fall Rubber Division meeting in the year they are selected, receiving up to $1,250 each. At a spring technical meeting, awardees are invited to attend on-site seminars and workshops, technical meetings, the Science and Technology Awards banquet, and the Sponsors' Reception. At a fall rubber expo and mini-expo, awardees are invited to attend on-site seminars and workshops, the Business Summit, technical sessions, and the Exhibitors' Reception as well as visit the exposition floor. The first four A. Wayne Place Memorial Award winners were Barry Deadman of American Biltrite, Jerry Fears of J. M. Clipper, Walter Hall of WLH-CTTRI, and Francis Kuebler of the Goodyear Tire & Rubber Company.

Also in 2003, the division established the first annual Student Colloquium, which is designed to provide a positive venue for students to present their work and meet with industry members in an encouraging and rewarding environment. The first colloquium was held on October 5, 2004, in Columbus, Ohio. The colloquia are open to both graduate and undergraduate students, and some can receive "initiatives" of a maximum five hundred dollars for their travel needs. The participants also have the opportunity to compete for four fifteen-hundred-dollar scholarships in the categories of Best Graduate Paper, Best Graduate Poster, Best Undergraduate Paper, and Best Undergraduate Poster. "However, the rewards for participation are not only monetary," said Miller. "This is an opportunity for students to present their best work to industry professionals and network with them during special events. In fact, several of our participants have received job offers on site!" Student colloquia also provide a forum for companies to showcase their internship and co-op programs during a job fair. Many companies participate and the program is growing. In addition, the division lists industry and academic internships on its web site, with Excel Polymers being the first company to participate.

While it does not participate as a sponsor, the division has supported scholarship efforts initiated by the local rubber groups and provides guidance when requested. In April 1954, the Akron Rubber Group, now known as the Ohio Rubber Group, was one of the first to establish a scholarship program by supporting "a total of four chemistry or chemical engineering students."[4] In 1992, the group handed over administration of the program to the University of Akron's Department of Development, where it became endowed.

The most recent element of the educational equation was the division's entree in 2003 to the world of a Monster.com-like, online career center. "This is the bottom line of our workforce recruitment and development initiative," said Miller. "Graduates need jobs, and our online Career Center provides employers a forum to advertise and for students to search for jobs. It's a good program that ties everything together, and it enables us to reach out and show them what is available in the industry. With the average age of a chemist around fifty-six years old, we need to address the new generation. How do we get the younger people into this industry? You have to get them involved from day one." Use of the center (http://careercenter.rubber.org), the only job site specifically targeting the rubber industry, is free to students and others wanting to post resumes or search for jobs.

One other instructional/educational element introduced by the division prior to Miller's arrival was the Rubber Recycling Topical Group (RRTG). Initiated in 1994, the RRTG is a nonprofit association of members dedicated to promoting and expanding rubber recycling. Its membership is comprised of processors of recycled postindustrial and postconsumer rubber (tires), consultants, raw feedstock suppliers, and end users of recycled rubber for rubber and plastic applications. The RRTG publishes the *Innovations in Recycling* newsletter quarterly and supports the division by providing educational and technical sessions on rubber recycling at the rubber expo shows.

Today, in addition to the undergraduate scholarships and rubber technology training awards, the division offers a variety of additional educational opportunities, including on-line courses, workshops, seminars, internships, co-ops, and career assistance through several associations, companies, and universities throughout the world.

BACK
INSIDE THE BOX

In every conceivable manner, the family is the
link to our past and bridge to our future.
ALEX HALEY, AUTHOR

Traveling primarily across three major interstate highways,
Lexington, Kentucky, is about five hours and three-hundred-
plus miles by car from Akron, Ohio. In 2001, Ed Miller
became acquainted with most of the potholes along that
route. That year, almost immediately after he became the
executive director of the Rubber Division, he multitasked
his way from Lexington to Akron in order to develop a
plan he believed would help create a more effective organization.

"The hardest part was my first four months on the
job," Miller said. "Because I was commuting every week

between Lexington and Akron. I would drive up on Sunday—I had an apartment in Akron—My family was in an apartment in Lexington while our home was being rebuilt, and I would put in, literally, eighteen- to twenty-hour days—I only got a few hours of sleep while I was working for the division. Then I would head back to Lexington on Thursday morning to see the progress of the rebuilding of our home."[1]

During that time, Miller developed a twenty-page "state of the division" report for the executive and steering committee members titled "Report of the Executive Director: The First Three Months." In the report, Miller analyzed the current status and relationship of the division with the American Chemical Society, the local rubber groups, personnel, infrastructure, and, of course, finances. In addition, he even enlisted the creative services of Katie Bruno, the talented daughter of Frank Kelley, to develop a new logo for the division. Miller envisioned a logo representative of the new Rubber Division and its positive future direction. Bruno's design was approved by both the ACS and the division steering committee. Her design is still used today, but Miller almost didn't make it past the three months he analyzed.

"I prepared my huge report, did a full analysis and presented it to the board in April of 2001," Miller said. "When I was done and asked for questions, the room was dead silent." In the excruciating pause of seemingly hours, Miller's future raced by him in nanoseconds. Then "three or four" bodacious souls fired verbal volleys. "They said, 'What? We're going to do what? What is this stuff?' I wasn't totally shocked," said Miller, "because I was expecting some kind of reaction, but I didn't hear anything positive." Miller left the room dejected, called his wife, Diane, back in Lexington, and told her in a very serious tone, "Don't sell the house. I think I'm coming home."

However, shortly afterward, one of the board members, Colleen McMahan, an executive with Advanced Elastomer Systems (AES), approached Miller and said, "Ed, don't get depressed. Don't quit. Remember, you just knocked our socks off. Give us time to evaluate this, talk about it and let it sink in. I'm convinced we're going to move forward, but have some confidence in us, too." An equally consoling Rudy School took Miller to dinner that night and convinced him to stay. Miller's only question to School in reply to his baptism by fire was,

"When you voted to go to a professional organization, did you really know what you voted for?" School's answer came within nine months. By that time, the steering and executive committees were all pulling in the same direction with Miller, and the change agents were no longer crawling—they were running.

Miller brought to the division a level and breadth of experience that hadn't existed with any predecessor who had responsibility for the organization. He had been a career U.S. Air Force engineer, graduating from the Air Force Academy with a B.S. in civil engineering, the University of New Mexico with a master's degree, and the University of Oklahoma with an M.B.A. In his twenty years with the military, he attained the rank of major, and during his career he taught graduate school and consulted in fifteen countries throughout Asia, Europe, the Middle East, and South America.

In 1992 he changed his flight plan. Shortly after he retired from the U.S. Air Force, Miller took a job as president and CEO of the Asphalt Institute based in Lexington, Kentucky. In that role he served as spokesperson for the asphalt producers, processors, marketing companies, and affiliated businesses within the oil industry. He also directed the institute's sixteen field offices in the United States and Canada, which included engineering, research, education, environmental, government affairs, and marketing operations. The assignment prepared him well for the challenges he would face in the Rubber Division.

"The Asphalt Institute is a technical association, and so is the Rubber Division," Miller said. "Even the words that [the division] used in their ad [seeking an executive director] fit almost exactly what I was doing with the Asphalt Institute. Everything just seemed to match." The steering and executive committees of the Rubber Division concurred. Miller was their man. "We wanted someone who could run the organization and be the CEO," said Rudy School, the Rubber Division chairman in 2001 who coordinated the hiring of Miller. "Once we hired Ed—and we knew we had the right individual—then we needed to address the other issues."[2]

While his hiring signaled the official beginning of the new Rubber Division, Ed Miller knew that he could not meet his own personal goals and the division's objectives by himself. He needed to organize a staff that was just as dedicated. While the first steps to complete the separation of university and division were taken, the new structure was still crawling and awaited its walk-definitive second steps, and they included assembling the staff and moving into yet another new office space. "We were located on the third floor of the Goodyear Polymer Center, with a small storage room on the fifth floor," Miller said. "We also had a storage room in Canal Place, which was filthy." Simultaneous to writing his eyebrow-raising epistle on the division's status, Miller also needed to determine which disciples were going to remain with the organization and begin the physical move to a different location—one where the division could finally establish its own autonomous identity. The one-million-dollar donation to the University of Akron by the Rubber Division that Chairman J. Marshall Dean publicly announced in 1997 would soon yield a rewarding dividend.

The construction of the new Polymer Engineering Academic Center on the northwest side of the campus was nearing completion, and the Rubber Division had cast a desiring eye toward the fourth floor of the structure. "Rudy School and I visited the new offices while they were still under construction, around March of 2001," said Miller. With a few minor alterations, they liked what they saw and went to Kent Marsden, the former division administrator who was still its landlord, to see if those changes could be made. Marsden handled the request, and Miller and a few staff members who would stay with the organization prepared to move.

Marsden was pleased with the transition period, in particular with the handling of the university employees. "With only one exception, all the employees either found jobs on campus or outside of the university," he said. "The executive committee of the Rubber Division had developed job descriptions and all the employees were encouraged to apply for the newly created positions."[3]

Between Miller and his infinitesimal but expanding team, the move to the new offices was completed in June 2001. "The limited staff we had did a great job on the move," he said. "[We] went to work going through everything in the old office and storage room . . . and got rid of what was no longer necessary. Only one other thing occurred. Sue Barr and I went to Canal Place and found mountains of outdated books, paperwork, et cetera, being stored—and being paid for. We spent a week going through all of the storage areas, throwing out items we no longer needed and packing up and moving other items—via our cars—to our new office storage area."

"It was a mess," Barr said of the Canal Place storage. She was one of the first three employees Miller hired. "We filled up many trash bags and several trash bins. We kept the important material and even stored some of it in the Lincoln Building (just to the northeast of the Polymer Engineering Academic Center)."[4] Barr was one of the university employees who, after a "reentry" interview process for those who wanted to leave the University of Akron's jurisdiction, elected to join the new Rubber Division organization. She became the expo and meeting sites manager, while Connie Morrison-Koons, the first person hired by Marge Bauer, was the meetings manager and Missy Benyon was the systems/interim education manager. Soon after Miller convinced the steering committee of the need—especially to prepare for the 2001 spring meeting in Providence, Rhode Island—he hired a few more employees. "We separated the systems and education positions, hired a new education manager, Vickie George, and a receptionist/membership manager, Tia Flammer, while making Missy (Beynon) full-time systems manager," Miller said. "Then we hired Gay Williams as meetings and technical programming manager when Connie decided to stay with the University."

When the initial hiring was complete, Miller distributed his time among completing his report and, literally, getting his family's house in order. The division-related tasks included "completely revising the budget line-item format to make it useful to the executive committee, developing internal policies and writing a complete employee handbook; working with the treasurer and division chair (Rudy School) to establish all benefit programs; putting in place the necessary insurance policies and finding an apartment and getting set up in it in Akron, while I continued to commute to Lexington to keep my family together and rebuild my fire-damaged home," he said. He added, "That was just the first four months."

During Miller's first hundred days, the division's new supporting structure began to materialize. To complete the initial staffing requirements, a finance manager, J. P. Schippert, was hired, and another receptionist, Dorothy Fiorella, replaced Flammer, who was quickly promoted to membership manager. The rest of the year Miller worked with his new staff and the various division committees to put new programs in place; worked with the staff and officers to reduce expenses and turn finances around; began development of the first true strategic plan, "and got up to speed on the American Chemical Society. It was a very busy and difficult but exciting year!" he said.

William J. Sparks, chairman of the Rubber Division in 1960 and president of the American Chemical Society in 1966, once defined the relationship between his two professional groups with tongue-in-cheek candor: "The Division of Rubber Chemistry is probably the most independent-minded, contrary—and lively—division in the ACS."[5] Ralph Graff, an emeritus member of both the American Chemical Society and the Rubber Division, was more diplomatic in his appraisal, and described the bond between the two groups as "one of the great balancing acts."

It was fairly common knowledge among the membership that there was a period of time when, like most consanguinities, the ACS and the Rubber Division had completely polarized opinions about the latter's operation as an autonomous entity. The actual historical precedent of the Rubber Division's formation underscores the very reasons behind its existence. From the birth of the India Rubber Section in 1909, and the transition to the Rubber Division in 1919, the organization had been overtly attempting to establish its own identity separate from the American Chemical Society. In sports vernacular, it wanted to run with its own ball. For the first sixty-five years, the ACS carefully scrutinized its divisions, but the independent-minded rubber chemists had other thoughts—though they would have to wait.

A definitive separation, of sorts, occurred in the mid-1970s. "The last time the Rubber Division met with the ACS was in 1976 in New York City, and then it wasn't until 1979 when we could have papers presented at our meetings from someone who wasn't a member of the ACS," Graff, the affable chairman of the Rubber Division in 1980, recalled. "Before that, anyone who gave a paper at the Rubber Division had to be a member of the American Chemical Society or have a co-author who was a member. We had people wanting to give papers . . . that we really wanted to hear. To make that happen, we'd have to run around and find somebody that was agreeable, and somebody that the author would agree to, and make him co-author so we could get the paper presented on the program."[6]

The all-time high number of papers presented during those years was 119 at the 1983 International Rubber Conference in Houston,[7] but the "Split of '76" and issues over what group had what hotel or how many rooms would be allocated to each, and so on, while seemingly trivial, weighed heavily on the volunteer structure of the organization and temporarily diverted their attention from the crux of their primary concerns—the financial bottom line.

Jim D'Ianni, because of his unique status as one of only three individuals to have held both the top elected positions in the American Chemical Society and the Rubber Division (Sparks was one, and the other was Harry L. Fisher, 1928 Rubber Division chairman and 1954 ACS president), also lent a somewhat bipartisan perspective to the relationship. As if to leave a verbal legacy, two months before D'Ianni departed this life on August 14, 2007, he shared, with characteristic good humor, some of his thoughts on many subjects, including the relationship between the Rubber Division and the ACS:[8]

> Well, the separation did not occur that long ago. I'd say it occurred because there was a mismatch between the attendance and the kind of papers that were being published. The American Chemical Society papers tended to be on fundamental work whereas the Rubber Division tended to emphasize applied chemistry, business activities, and other things . . . other than plain chemistry. Things today are so complicated you get many different fields where people have to work together in order to solve a problem. They can't do it by themselves. So, it's changed. . . .
> It's a living, dynamic organization, and we've got some top notch people in Washington right now running the American Chemical Society, including the CEO Madeleine Jacobs; and the president is a woman. The past president is a woman, and the vice president is a woman.[9] The women are taking over.

D'Ianni didn't mention Barbara Hodsdon, part of the American Chemical Society's highly competent and effective gynecocracy, whose official title was manager of divisional activities and later department head of meetings of the ACS. In a sense, she was the councilor to the Rubber Division. "She was the best staff person I've ever seen," said Charles Rader from Monsanto, who was Rubber Division chairman in 1986. "She is a very classy person and was a great deal of help to us during her tenure."[10]

Hodsdon was the ACS's representative to the division from the mid-1970s until her retirement in 1989. "I was liaison between the national staff and the division," Hodsdon (now Barbara Ullyot) recalled:

> The Rubber Division was unique. The big difference between them and the other divisions of the ACS was that they always held separate meetings whereas the other divisions met as part of our [ACS] spring and fall national meetings. The Rubber Division felt that its interests were such that they were better meeting entirely separately. I went to each of the division's two national meetings each year and I was warmly received and genuinely welcomed. I had the unique position that I met with the men in the business end of things and also met with the wives on the more social end of things. I also maintained constant contact with their councilors.[11]

D'Ianni concurred with Hodsdon's appraisal of the one-on-one relationship of ACS personnel and the Rubber Division. Even in his health-plagued last years, ever the optimist and upbeat individual, D'Ianni had a vision for his fellow rubber chemists and knew that past issues were just opportunities for future growth. "Our future is bright, because it is challenging. I have attempted to do what science is expected to do, and this is to look into the future and predict what it may hold. To say that our prospects are exciting is an understatement indeed."[12]

D'Ianni was the holder or coholder of seventeen patents and the author or coauthor of more than fifty scientific articles. He took a leave of absence from Goodyear in 1946 to serve as chief of the polymer research branch of the Office of Rubber Reserve in Washington, D.C. In 1977, he received the division's highest honor, the Charles Goodyear Medal.

Susan E. Calvo, a University of Akron administrator in the college of arts and sciences, spent a great deal of time with D'Ianni and compiled a book of his papers, memoirs, and so on. D'Ianni said of Susan, "She's a very nice person, but very enthusiastic about what she wrote." Then he offered a suggestion: "Where you see the hyperbole of my efforts, please tone it down, won't you?" D'Ianni was ninety-three-years wise.

After his analysis of the relationship with the ACS and many issues held over from the previous structure—including several concerns of historical spam—Ed Miller realized that, in order to rejoin the bonds, it would take a Herculean effort on his and the division chair's part—and time. His intent wasn't to change the division's unique status as a special and separate entity of the ACS and jeopardize its own established identity, but he really believed "that strategic partnerships could help further the overall industry goals such as the need to educate the new generation of chemists and technicians."

In June 2001, in one of the first official moves from his new coordinates, Miller and Rubber Division chairman Rudy School visited an old friend in Washington, D.C.—the American Chemical Society—to meet with some "key staff." Their initial visit did not include the executive director because of a transition in that position. "We opened the door," Miller said. "We started realizing what the ACS really is and what they have to offer that [we] could benefit from." The trip, in part, was Miller's self-reaffirmation of the need to rekindle the relationship with the ACS, and to extend a partnering hand to other similar groups. After many years of distancing itself from its parent society, the Rubber Division, like a prodigal son, reached out and returned to the fold, if only to establish a periodic pilgrimage. "For years we had been fairly autonomous," said Miller. "Really, we were operating out of the box of the ACS. In my opinion, we really didn't have enough direct contact with the ACS. I think we had lost that through the years."

Madeleine Jacobs is a very savvy businesswoman. She has an undergraduate degree in chemistry from George Washington University, completed course work for a master's degree in organic chemistry at the University of Maryland, and received an honorary doctor of science degree from her alma mater, George Washington, in 2003. Jacobs is a multi-honored science journalist who became intimately familiar with the chemical industry when she served for eight and a half years as editor-in-chief and two years as managing editor of *Chemical & Engi-*

neering News, the weekly news magazine of the chemical world published by the American Chemical Society. In 2004, she accepted the position of executive director and CEO of the ACS, where she has shared her extensive familiarity and understanding of the society's programs, products, and services. Jacobs was a strategic contact for Miller. "We established and built a rapport with her," he said. "And we invited her to key events for our meetings, and she invited the [Rubber Division] chair and I to some of her meetings. So we started to build a good relationship and to share ideas."

The division even reached out to both the Akron Section of the American Chemical Society and the ACS's Division of Polymer Chemistry. "We had a joint technical luncheon with the ACS Akron Local Section at our spring meeting in 2006," said Miller. "Then we had a joint technical symposium with the ACS Division of Polymer Chemistry at our fall meeting in 2007." The fall 2007 event also was attended by ACS's president, Catherine T. Hunt, who served as a judge for the student colloquium on the heels of her predecessor, Ann Nally's, presence and participation at the 2006 fall meetings.

The ACS's top executive believes the meetings with Miller, his staff, and the society's elected chairs have been mutually beneficial. "When we note the current respective strategic plans, we see how the Rubber Division and the American Chemical Society at large are partners within the chemical enterprise," Jacobs said. "We both seek to serve as a global resource in education, science, and technology information and programs; have highly qualified and committed volunteers, and both plans reflect the growing trend of globalization and the passion for chemistry by their goals for research, education, and innovation."[13]

Today, Miller and the division chair and chair-elect meet every summer at ACS headquarters in Washington to have lunch or dinner and discuss increased use of ACS services by the division along with idea sharing and other topics of mutual concern.

———————

By the end of July 2001, and after his initial meetings with the American Chemical society, Miller and his family completed their move from Kentucky to the Akron area and were now comfortable in their new Ohio home. For the next three intensive years, he devoted his waking and often unconscious minutes to the implementation of the division's strategic plan.

During those first one thousand days, Miller and the new staff worked closely with the executive committee and other volunteers to put the professional association's organizational structure in place. They needed to introduce new online technology, strategic planning with associated business plans, formal committee charters, and daily financial accountability; initiate the student colloquium, career center, and formal training of new officers, directors, and rubber group chairs. Finally, they had to institute the Rubber Modified Asphalt Conference, working with Michael Blumenthal, president of the Rubber Manufacturer's Association, Doug Carlson of the Rubber Pavements Association, and three Asphalt Association presidents.

"We had started out the transition with one university employee, Joan Long, as the librarian," said Miller. "When Joan retired at the end of 2001, we hired Chris Laursen." In addition, by 2004 Miller had completed the first of his "initial" staff additions with the hiring of a marketing and meetings manager, Tammy Sobleskie. In 2009, the staff included Systems Manager Missy Beynon, Expo and Future Sites Manager Sue Barr, Marketing Manager Karen May, Education and Publications Manager Christie Robinson, Membership and Technical Programming Manager Linda McClure, Accounting Manager Lakisha Miller, and Meeting and Office Manager Melanie Avdeyev.

Miller, his staff, and key member volunteers followed the first few years' achievements with the Thermoplastic Elastomer (TPE) Conference in 2006, reaffirmation of such solid programs as the Business Summit, increased support of local rubber groups, global affiliations through the new Partnership Development Committee chaired by John Long, and more focused marketing efforts. "In 2001 we had made a commitment to extend all of our services via our web site," he said. "We wanted everything to be online." Taking a page out of the ACS information technology manual, the Rubber Division placed its membership directory online and began to use its web site for electronic voting. Using input from the Local Rubber Groups and the efforts of Systems Manager Missy Beynon and Membership Manager Linda McClure, they expanded online services to these groups to include a Technical Speakers Clearinghouse, newsletter, and a new single membership application and renewal form. "It was something the local groups wanted," Miller said. "They needed a single application membership form instead of having to fill one out for each Rubber Group in addition to the division. Now members can renew all of their memberships using one form and one payment. In addition, each of the participating groups has access to a pass code and can make their own changes to their sites. So we provided them with a content-managed 'one-stop shopping' option."

While a great deal has been accomplished in the division's first century, Miller says, "there's plenty more that we still need to do." Miller recognizes his responsibility of providing the daily operational leadership as well as serving as a spokesperson, developing partnerships with other organizations, and maintaining good communications with the governing committees, rubber groups, and the American Chemical Society.

"It's a responsibility to the membership and volunteers," he said. "We must constantly grow to meet today's needs and tomorrow's requirements." Recognizing the current and future needs of the rubber industry today, the division has embraced a new vision: To enhance science, technology, and business across the evolving elastomeric community. "With this, the Division will take a lead role in expanding the elastomeric profession and individual development through educational, technical and interactive activities," Miller added.

The Rubber Division has begun its second century much like it did the first—in a continuous stage of development and growth. While the division's growth curve over the first one hundred years might resemble a graph of the heights of daily solar flares, those peaks and valleys have provided the organization with continuous opportunities to reassess its value to members whose industries are themselves constantly changing.

High technology in every phase of the business continues to evolve, grow, and rule over the traditional blood-sweat-and-tears-based businesses, and the opportunities of the early 1900s—and even of the twentieth century—are different from those the division faces in the twenty-first century. Yet an equivocal amount of those formative years' pioneering spirit continues to permeate the new organization. To that principle, Miller added, "We constantly remind each person on our staff that this association only exists *because* of our membership and *volunteers*. Having a professional staff never takes away from that. The volunteers are still the lifelines of our organization. We keep that in mind daily . . . and live by it."

APPENDIX 1

OFFICERS OF THE RUBBER DIVISION, 1909–2009

YEAR	CHAIRMAN	SECRETARY*	TREASURER
1909	Charles C. Goodrich, B. F. Goodrich	Frederick J. Maywald	
1910	Charles C. Goodrich, B. F. Goodrich	Frederick J. Maywald	
1911	Charles C. Goodrich, B. F. Goodrich	Frederick J. Maywald	
1912	David A. Cutler, Alfred Hale Rubber	Dorris Whipple, Standard Chemical	
1913	David A. Cutler, Alfred Hale Rubber	Dorris Whipple, Standard Chemical	
1914	David A. Cutler, Alfred Hale Rubber	Dorris Whipple, Standard Chemical	
1915		—	
1916	Lothar E. Weber	John B. Tuttle, Firestone	
1917	Lothar E. Weber	John B. Tuttle, Firestone	
1918	Lothar E. Weber	John B. Tuttle, Firestone	
1919	John B. Tuttle, Firestone	Arnold H. Smith, U.S. Bureau of Standards/Goodyear	
1920	Warren K. Lewis, Massachusetts Institute of Technology	Arnold H. Smith, U.S. Bureau of Standards/Goodyear	
1921	Walter W. Evans, Goodyear	Arnold H. Smith, Thermoid	
1922	Clayton Wing Bedford, Goodyear/ Rubber Service Labs/Goodrich	Arnold H. Smith, Rubber Service Laboratories	
1923	William B. Weigand, Ames Holden	Arnold H. Smith, Rubber Service Laboratories	
1924	Elwood B. Spear, Goodyear	Arnold H. Smith, Rubber Service Laboratories	
1925	Charles R. Boggs, Simplex Wire and Cable	Arnold H. Smith, Rubber Service Laboratories	
1926	John M. Bierer, Boston Woven Hose and Rubber	Arnold H. Smith, Rubber Service Laboratories	
1927	Raymond P. Dinsmore, Goodyear	Arnold H. Smith, Rubber Service Laboratories	
1928	Harry L. Fisher, U.S. Rubber	Hezzelton E. Simmons, University of Akron	
1929	Arnold H. Smith, Rubber Service Laboratories	Hezzelton E. Simmons, University of Akron	
1930	Stanley Krall, Fisk Rubber	Hezzelton E. Simmons, University of Akron	
1931	Herbert A. Winkelmann, General Atlas Carbon	Hezzelton E. Simmons, University of Akron	

continued

YEAR	CHAIRMAN	SECRETARY*	TREASURER
1932	Ernest R. Bridgewater, DuPont	Hezzelton E. Simmons, University of Akron	
1933	Lorin B. Sebrell, Goodyear	Hezzelton E. Simmons, University of Akron	
1934	Ira Williams, DuPont	Hezzelton E. Simmons, University of Akron	
1935	Sidney M. Caldwell, U.S. Rubber	Chester W. Christensen, Rubber Service Laboratories	
1936	Norman A. Shepard, Firestone	Chester W. Christensen, Rubber Service Laboratories	
1937	Harlan L. Trumbull, B. F. Goodrich	Chester W. Christensen, Monsanto	
1938	Archie R. Kemp, Bell Telephone Laboratories	Chester W. Christensen, Monsanto	
1939	George K. Hinshaw, Goodyear	Chester W. Christensen, Monsanto	Chester W. Christensen, Monsanto
1940	Ernest B. Curtis, R. T. Vanderbilt	Howard I Cramer, University of Akron;	Chester W. Christensen, Monsanto
1941	Roscoe H. Gerke, U.S. Rubber	Howard I Cramer, University of Akron;	Chester W. Christensen, Monsanto
1942	John N. Street, Firestone	Howard I Cramer, Sharples Chemicals;	Chester W. Christensen, Monsanto
1943	John T. Blake, Simplex Wire and Cable	Howard I Cramer, Sharples Chemicals;	Chester W. Christensen, Monsanto
1944	Harold Gray, B. F. Goodrich	Howard I Cramer, Sharples Chemicals;	Chester W. Christensen, Monsanto
1945	Willis A. Gibbons, U.S. Rubber	Howard I Cramer, Sharples Chemicals;	Chester W. Christensen, Monsanto
1946	Willis A. Gibbons, U.S. Rubber	Howard I Cramer, Sharples Chemicals;	Chester W. Christensen, Monsanto
1947	Walter W. Vogt, Goodyear	Charles R. Haynes, Binney and Smith;	Chester W. Christensen, Monsanto
1948	Harry E. Outcault, St. Joseph Lead	Charles R. Haynes, Binney and Smith;	Chester W. Christensen, Monsanto
1949	Howard I Cramer, Sharples Chemicals	Charles R. Haynes, Binney and Smith;	Chester W. Christensen, Monsanto
1950	Frederick W. Stavely, Firestone	Charles R. Haynes, Binney and Smith;	Chester W. Christensen, Monsanto
1951	John H. Fielding, Armstrong	Charles R. Haynes, Binney and Smith;	Chester W. Christensen, Monsanto
1952	Waldo L. Semon, B. F. Goodrich	Charles R. Haynes, Binney and Smith;	Amos W. Oakleaf, Phillips Chemical
1953	Seward G. Byam, DuPont	Charles R. Haynes, Binney and Smith;	Amos W. Oakleaf, Phillips Chemical
1954	James C. Walton, Boston Woven Hose and Rubber	Arthur M. Neal, DuPont	Amos W. Oakleaf, Phillips Chemical
1955	John M. Ball, Midwest Rubber Reclaiming	Arthur M. Neal, DuPont	Amos W. Oakleaf, Phillips Chemical
1956	Arthur E. Juve, B. F. Goodrich	Arthur M. Neal, DuPont	George E. Popp, Phillips Chemical
1957	Benjamin S. Garvey Jr., Penn Salt Manufacturing	Arthur M. Neal, DuPont	George E. Popp, Phillips Chemical

YEAR	CHAIRMAN	SECRETARY*	TREASURER
1958	Raymond F. Dunbrook, Firestone	Roscoe H. Gerke, U.S. Rubber	George E. Popp, Phillips Chemical
1959	Emil H. Krismann, DuPont	Roscoe H. Gerke, U.S. Rubber	George E. Popp, Phillips Chemical
1960	William J. Sparks, Esso Research and Engineering	Roscoe H. Gerke, U.S. Rubber	George E. Popp, Phillips Chemical
1961	Wesley S. Coe, Naugatuck Chemical Div. of U.S. Rubber	Louis H. Howland, Naugatuck Div. of U.S. Rubber	Dale F. Behney, Harwick Standard Chemical H.I.
1962	George E. Popp, Phillips Chemical	Louis H. Howland, Naugatuck Div., U.S. Rubber	Dale F. Behney, Harwick Standard Chemical
1963	Gilbert H. Swart, General Tire	Louis H. Howland, Naugatuck Div., U.S. Rubber	Dale F. Behney, Harwick Standard Chemical
1964	James D. D'Ianni, Goodyear	George N. Vacca, Bell Telephone Laboratories	John H. Gifford, Witco Chemical
1965	Edwin B. Newton, B. F. Goodrich	George N. Vacca, Bell Telephone Laboratories	John H. Gifford, Witco Chemical
1966	Norman S. Grace, Dunlop Research Center	George N. Vacca, Bell Telephone Laboratories	John H. Gifford, Witco Chemical
1967	Dale F. Behney, Harwick Standard Chemical	George C. Winspear, R. T. Vanderbilt	Richard A. Garrett, Armstrong Cork
1968	Glen Alliger, Firestone	George C. Winspear, R. T. Vanderbilt	Richard A. Garrett, Armstrong Cork
1969	Thomas H. Rogers, Goodyear	George C. Winspear, R. T. Vanderbilt	Richard A. Garrett, Armstrong Cork
1970	Paul G. Roach, Texas-U.S. Chem	Francis M. O'Connor, Harwick Chemical	Joseph C. Ambelang, Goodyear
1971	John H. Gifford, Continental Carbon	Francis M. O'Connor, Harwick Chemical	Joseph C. Ambelang, Goodyear
1972	Albert E. Laurence, R. T. Vanderbilt	Francis M. O'Connor, Harwick Chemical	Joseph C. Ambelang, Goodyear
1973	Eli M. Dannenberg, Cabot	H. Webster Day, DuPont	Ralph S. Graff, DuPont
1974	Francis M. O'Connor, Harwick Chemical	H. Webster Day, DuPont	Ralph S. Graff, DuPont
1975	Ben Kastein, Firestone	Walter C. Rowe, Firestone	Ralph S. Graff, DuPont
1976	Earle C. Gregg Jr., B. F. Goodrich	Walter C. Rowe, Firestone	Daniel A. Meyer, General Tire
1977	H. Webster Day, DuPont	Donald W. Gorman, R. T. Vanderbilt	Daniel A. Meyer, General Tire
1978	Frank B. Floren, Exxon Chemical	Donald W. Gorman, R. T. Vanderbilt	Daniel A. Meyer, General Tire
1979	Ralph F. Anderson, B. F. Goodrich	Donald W. Gorman, R. T. Vanderbilt	Harold J. Herzlich, Armstrong
1980	Ralph S. Graff, DuPont	Thomas Jones, Wyrough and Loser	Harold J. Herzlich, Armstrong
1981	Donald W. Gorman, R. T. Vanderbilt	Thomas Jones, Wyrough and Loser	Lloyd D. Treleaven, Columbian
1982	Harold J. Herzlich, Armstrong	Eldon R. Sourwine, Firestone	Lloyd D. Treleaven, Columbian
1983	Thomas N. Loser, Wyrough and Loser	Eldon R. Sourwine, Firestone	Charles P. Rader, Monsanto
1984	Lloyd D. Treleaven, Columbian	Ronald J. Ohm, ASARCO	Charles P. Rader, Monsanto
1985	Eldon R. Sourwine, Firestone	Ronald J. Ohm, ASARCO	Charlotte Sauer, 3M

YEAR	CHAIRMAN	SECRETARY*	TREASURER
1986	Charles P. Rader, Monsanto	John W. Messner, Concarb-Witco	Charlotte Sauer, 3M
1987	Thomas Jones, Wyrough and Loser	John W. Messner, Concarb-Witco	Peter W. Spink, Monsanto
1988	Robert A. Pett, Ford	Albert J. Brandau, DuPont	Peter W. Spink, Monsanto
1989	John W. Messner, Concarb-Witco	Albert J. Brandau, DuPont	Daniel L. Hertz, Seals Eastern
1990	Peter W. Spink, Monsanto	Roger K. Price, R. T. Vanderbilt	Daniel L. Hertz, Seals Eastern
1991	Albert J. Brandau, DuPont	Roger K. Price, R. T. Vanderbilt	Thomas J. Dendinger, Cooper Engineered Products Division
1992	William J. Hines, Lord Corp.	John M. Long, Uniroyal Goodrich Tire	Thomas J. Dendinger, Cooper Engineered Products Division
1993	Roger K. Price, R. T. Vanderbilt	John M. Long, Uniroyal Goodrich Tire	Patrick J. Heitz, Barbe America
1994	Thomas J. Dendinger, Cooper Engineered Products Division	J. Marshall Dean III, Harwick Chemical	Patrick J. Heitz, Barbe America
1995	John M. Long, DSM Copolymer	J. Marshall Dean III, Harwick Chemical	Stanley M. Mezynski, Goodyear Tire and Rubber
1996	Daniel J. Hertz Jr., Seals Eastern	Donald E. Mackey, Zeon Chemicals	Stanley M. Mezynski, Goodyear Tire and Rubber
1997	J. Marshall Dean III, Harwick Chemical	Donald E. Mackey, Zeon Chemicals	Toms B. Royal, H. M. Royal
1998	Stanley M. Mezynski, Goodyear Tire and Rubber	Rudy J. School, R. T. Vanderbilt	Toms B. Royal, H. M. Royal
1999	Donald E. Mackey, Zeon Chemicals	Rudy J. School, R. T. Vanderbilt	Christopher G. Probasco, Flexsys America
2000	Toms B. Royal, H. M. Royal	Ernie L. Puskas Jr., Littlern Corp.	Christopher G. Probasco, Flexsys America
2001	Rudy J. School, R. T. Vanderbilt	Mark A. Petras, ChemRep	Kurt S. Nygaard, Harwick Standard
2002	Richard J. Hupp, Enichem America	David R. O'Brien, PolyOne	Paul V. Esposito, Uniroyal Chemical
2003	Mark Petras, ChemRep	David R. O'Brien, PolyOne	P. Andrew Claytor, Teknor Apex
2004	Paul V. Esposito, Uniroyal Chemical	John Boruta, Rhein Chemie	P. Andrew Claytor, Teknor Apex
2005	David R. O'Brien, PolyOne	John Boruta, Rhein Chemie	Allan D. Feit, Goodyear
2006	P. Andrew Claytor, Teknor Apex	Gary Horning, Polymer Valley Chemicals	Allan D. Feit, Goodyear
2007	John Boruta, Consultant	Gary Horning, Polymer Valley Chemicals	Timothy S. Dickerson, R. T. Vanderbilt
2008	Allan D. Feit, Consultant	Walter H. Waddell, ExxonMobil Chemical	Timothy S. Dickerson, R. T. Vanderbilt
2009	Gary Horning, Polymer Valley Chemicals	Walter H. Waddell, ExxonMobil Chemical	Joseph Walker, Freudenberg-NOK General Partnership

*Up until 1940, there were only two officers (Chair and Secretary-Treasurer). Starting in 1940, the positions were split into three (Chair, Secretary, and Treasurer).

APPENDIX 2

EDITORS OF *RUBBER CHEMISTRY AND TECHNOLOGY*, 1928–2009

DATES	EDITOR
1928–1957	Carroll C. Davis, Boston Woven Hose and Rubber
1958–1964	David Craig, B. F. Goodrich
1964–1968	Edward M. Bevilacqua, Uniroyal
1969–1974	Earl C. Gregg Jr., B. F. Goodrich
1975–1977	Hans K. Frensdorff, DuPont
1977–1983	Aubert Y. Coran, Monsanto
1984–1990	Gary R. Hamed, University of Akron
1991–1999	C. Michael Roland, Naval Research Laboratory
2000–2001	Frederick Ignatz-Hoover, Flexsys America LP
2002–present	Krishna C. Baranwal, Akron Rubber Development Laboratory (retired 2006)

APPENDIX 3

CHARLES GOODYEAR MEDALISTS

The purpose of the Charles Goodyear Medal is to perpetuate the memory of Charles Goodyear as the discoverer of the vulcanization of rubber by honoring individuals for outstanding invention, innovation, or development that has resulted in a significant change or contribution to the nature of the rubber industry.

The award consists of six thousand dollars, a gold medal, a framed certificate, gratis lifetime Rubber Division affiliate membership, and one thousand dollars for expenses incurred incidental to attending the awards ceremony, as well as two nights complimentary lodging at the division meeting. The recipient is expected to deliver a lecture covering the background, development, implementation, and commercialization of the invention or innovation pertinent to the award.

The award was established by the Rubber Division in 1941. It is supported solely by the Rubber Division and is open and not restricted to Rubber Division membership.

2009	James L. White
2008	Joseph P. Kennedy
2007	Karl-Alfred Grosch
2006	Robert F. Landel
2003	Graham J. Lake

2001	Yasuyuki Tanaka
2000	Jack L. Koenig
1999	James E. Mark
1998	Jean-Baptiste Donnet
1997	Adel F. Halasa
1996	Siegfried Wolff
1995	Aubert Y. Coran
1994	Alan G. Thomas
1993	Leo Mandelkern
1992	Ronald S. Rivlin
1991	Edwin J. Vandenberg
1990	Alan N. Gent
1989	Jean-Marie Massoubre
1988	Herman F. Mark
1987	Norman R. Legge
1986	Leonard Mullins
1985	Maurice Morton
1984	Herman E. Schroeder
1983	J. Reid Shelton
1982	Adolf Schallamach
1981	John D. Ferry
1980	Samuel E. Horne Jr.
1979	Francis P. Baldwin
1978	Frank Herzegh
1977	James D. D'Ianni
1976	Earl Warrick
1975	Otto Bayer
1974	Joe C. Krejci
1973	Arnold M. Collins
1972	Frederick W. Stavely
1971	Harold J. Osterhof
1970	Samuel D. Gehman
1969	Robert M. Thomas
1968	Paul J. Flory

1967	Norman Bekkedahl
1966	Edward A. Murphy
1965	Benjamin S. Garvey
1964	Arthur E. Juve
1963	William J. Sparks
1962	Melvin Mooney
1961	Herbert A. Winkelmann
1960	William B. Weigand
1959	Fernley H. Banbury
1958	Joseph C. Patrick
1957	Arthur W. Carpenter
1956	Sidney M. Caldwell
1955	Raymond P. Dinsmore
1954	George S. Whitby
1953	John T. Blake
1952	Hezzelton E. Simmons
1951	William C. Geer
1950	Carroll C. Davis
1949	Harry L. Fisher
1948	George Oenslager
1946	Ira Williams
1944	Waldo L. Semon
1942	Lorin B. Sebrell
1941	David Spence

MELVIN MOONEY DISTINGUISHED TECHNOLOGY AWARD

The purpose of the Melvin Mooney Distinguished Technology Award is to perpetuate the memory of Melvin Mooney, the developer of the Mooney viscometer and other testing equipment, by honoring Rubber Division members or affiliate members who have exhibited exceptional technical competency by making significant and repeated contributions to rubber science and technology.

This award consists of three thousand dollars, an engraved plaque, and gratis lifetime Rubber Division affiliate membership. The recipient also receives five hundred dollars for travel expenses incurred in attending the awards ceremony meeting.

The award was established by the Rubber Division in 1982. In 2007, Lion Copolymer assumed sponsorship. The award is restricted to Rubber Division members and affiliate members.

Year	Recipient
2009	Frederick Ignatz-Hoover
2008	Robert P. Lattimer
2007	Daniel L. Hertz Jr.
2006	Meng-Jiao Wang
2005	Kenneth F. Castner
2004	Oon Hock Yeoh
2003	Walter H. Waddell
2002	C. Michael Roland
2000	Joseph Kuczkowski
1999	Avraam I. Isayev
1998	Henry Hsieh
1997	Russell A. Livigni
1995	Edward N. Kresge
1994	Noboru Tokita
1993	John R. Dunn
1992	Robert W. Layer
1991	Charles S. Schollenberger
1990	Gerard Kraus
1989	Joginder Lal
1988	John G. Sommer
1987	Avrom I. Medalia
1986	Albert M. Gessler
1985	William M. Hess
1984	Eli M. Dannenberg
1983	Aubert Y. Coran
1982	J. Roger Beatty

CHEMISTRY OF THERMOPLASTIC ELASTOMERS AWARD

The Chemistry of Thermoplastic Elastomers Award honors significant contributions to the advancement of the chemistry of thermoplastic elastomers. This award includes four thousand dollars, an engraved plaque, and five hundred dollars for travel expenses incurred in attendance at the awards ceremony meeting.

The award was established by the Rubber Division in 1991 as a part of its continuing effort to recognize the contributions of scientists in the field of thermoplastic elastomers. In 2009, past chair Ralph Graff assumed sponsorship. Particular emphasis is placed on innovations that have yielded significant new commercial or patentable materials. Patentable innovations in process chemistry for the production of new thermoplastic elastomers will also be eligible for this award. This award is open and not restricted to Rubber Division membership.

2009	Judit E. Puskas
2008	Richard J. Spontak
2007	Dale J. Meier
2006	Garth L. Wilkes
2002	Anil K. Bhowmick
2001	James McGrath
1996	Charles S. Schollenberger
1993	Geoffrey Holden
1992	William K. Witsiepe
1991	Aubert Y. Coran and Raman P. Patel

FERNLEY H. BANBURY AWARD

The Fernley H. Banbury Award perpetuates the memory of Fernley H. Banbury, the inventor and developer of the internal mixer that bears his name, by honoring innovations of production equipment widely used in the manufacture of rubber or rubber-like articles of importance.

The award consists of three thousand dollars, an engraved plaque, and five hundred dollars for travel expenses incurred in attending the awards ceremony meeting. This award was established by the Rubber

Division in 1986 as a part of its continuing effort to recognize the contributions of scientists and engineers in developing production equipment, control systems, and instrumentation widely used in the manufacture of rubber or rubber-like articles on importance. This award is currently sponsored by the Farrel Company. It is open and not restricted to Rubber Division membership.

2009	Donald J. Plazek
2006	Christopher W. Macosko
2005	Renato Caretta
2003	Bryan Willoughby
2002	William F. Watson
1993	Neland Onstott
1989	John P. Porter
1988	Engelbert G. Harms
1987	R. Warren Wise

GEORGE STAFFORD WHITBY AWARD

The George Stafford Whitby Award perpetuates the memory of George S. Whitby, head of the rubber laboratory at the University of Akron and for years the only one who taught rubber chemistry in the United States, by honoring outstanding international teachers of chemistry and polymer science and recognizing innovative research. The award consists of three thousand dollars, an engraved plaque, and five hundred dollars for travel expenses incurred in attending the awards ceremony meeting.

This award was established by the Rubber Division in 1986 as a part of its continuing effort to honor teachers and academic scientists for distinguished innovative and inspirational teaching and research in chemistry and polymer science. The award is currently sponsored by Cabot Corporation.

The activities recognized by the awards may lie in the fields of teaching (at any level), organization and administration, influential writing, standards of instruction, and public enlightenment. This award is open and not restricted to Rubber Division membership.

2009	James E. McGrath
2008	Frank N. Kelley
2007	Burak Erman
2006	Gregory G. McKenna
2005	Richard J. Farris
2004	Roderick P. Quirk
2003	Sadhan K. De
2002	Liliane Bokobza
2001	Gary Hamed
1998	Jean-Maurice Vergnaud
1997	Anil K. Bhowmick
1996	Joseph P. Kennedy
1995	Harry L. Frisch
1994	Jack L. Koenig
1993	Donald J. Plazek
1992	Raymond B. Seymour
1991	James E. Mark
1990	Howard L. Stephens
1989	Jean-Baptiste Donnet
1988	Leo Mandelkern
1987	Alan N. Gent

SPARKS-THOMAS AWARD

The purpose of the Sparks-Thomas Award is to perpetuate the memory of William J. Sparks and Robert M. Thomas, chemists who developed butyl rubber by recognizing and encouraging outstanding scientific contributions and innovations in the field of elastomers by younger scientists, technologists, and engineers. The award consists of four thousand dollars, an engraved plaque, and five hundred dollars for travel expenses incurred in attending the award ceremony meeting.

The award was established in 1986 and is supported by the Exxon-Mobil Chemical Company. Special consideration may be given to areas that have not been recognized recently. Recognition will also be given to originality and independence of thought and to the technological impact of the nominee's contribution. The nominee may be a citizen of any country and must be within twenty-five years of earning a baccalaureate degree. This award is open and not restricted to Rubber Division membership.

2009	John M. Baldwin
2008	Christopher G. Robertson
2007	William V. Mars
2006	Vassilios Galiatsatos
2005	Mark D. Foster
2004	Andy H. Tsou
1998	Anthony J. Dias
1997	Maria D. Ellul
1993	Walter H. Waddell
1991	C. Michael Roland
1990	Robert P. Lattimer
1987	Gary R. Hamed

ARNOLD SMITH SPECIAL SERVICE AWARD

Recipients of the Arnold Smith Special Service Award are chosen based on years and type of service to the Rubber Division. Recipients receive a plaque designating them as the award winner.

2008	Leo Whalen
2004	John Byers
1996	Walt Warner
1995	Wesley Whittington
1995	Bob Klingender
1994	Ed Kresge
1993	Charlotte Sauer
1992	Paul Graham
1991	Ron Ohm

DISTINGUISHED SERVICE AWARD

Recipients of the Distinguished Service Award are chosen based on having significantly exceeded the minimum requirements for the Arnold Smith Special Service Award. Recipients receive a plaque, a certificate, and a special gift from the Rubber Division.

2008	Richard J. Hupp
2007	Rudy School
2006	Peter Spink
2005	Roger Price
2005	Walter Waddell
2004	Chris Probasco
2003	John Messner
2003	Bill Klingensmith
2002	Russ Mazzeo
2001	Kris Baranwal
2001	Lyle Ryder
2000	Dan Hertz
1998	Don Gorman
1993	Jim D'Ianni
1991	Ralph Graff
1990	Francis O'Conner
1989	Howard Stephens
1988	Ben Kastein
1985	Ernest Zielasko
1982	Marge Bauer
1980	Jack Carr
1975	Ralph Wolf
1974	Earl Grey
1973	Mel Lerner

CERTIFICATE OF SPECIAL APPRECIATION

Recipients of the Certificate of Special Appreciation are those individuals who have made a significant contribution in the form of a single activity or series of related activities. Recipients receive a certificate.

2007	Dave Paulin
2005	Vipin Kothari
1997	Job Lippincott
1984	Peter Yurcick
1984	Bob Pett
1982	Howard Stephens
1981	Albert Lawrence
1981	Richard Edwards
1981	A. Durwin Dingle
1981	Albert Gessler
1980	Henry Remsberg
1979	A. H. Woodard
1979	Merton Studebaker
1979	J. Roger Beatty
1978	H. Carl Fernsdorff
1978	Kenneth Hacker
1978	Herbert Due
1978	Warren Carter
1978	Ralph Weaton
1977	Elastomerics
1977	*Rubber World*
1977	*Rubber & Plastics News*
1976	Floyd Conant
1976	Lee Borah
1976	Herbert Endres
1974	Cap V. Lundberg
1973	L. D. Loan
1973	R. J. Janssen

INTERNATIONAL RUBBER SCIENCE HALL OF FAME

The International Rubber Science Hall of Fame (IRSHF) was established in 1958 on the occasion of the celebration of the fiftieth anniversary of the start of the first course in rubber chemistry at Buchtel College in Akron, Ohio. Selection is limited to those who made a substantial contribution to the understanding of rubber-like materials or who were responsible for an outstanding invention, and only past contributors are eligible. The mechanism for the selection process is a joint committee from the University of Akron and the Rubber Division.

2008	Ronald S. Rivlin
2007	Ralph Milkovich
2006	David Tabor
2005	Thor L. Smith
2004	Frederic Stanley Kipping
2003	John D. Ferry
2002	Michael Szwarc
2001	Leonard Mullins
2000	Wendell V. Smith
1999	William J. Sparks and Robert M. Thomas
1998	Adolf Schallamach
1996	Gerard Kraus
1995	Lawrence A. Wood
1994	Maurice Morton
1993	Herman F. Mark
1992	Philip Schidrowitz
1991	Jean Le Bras
1990	George V. Vinogradov
1989	Merton L. Studebaker
1988	Edward M. Bevilacqua
1987	L. R. G. Treloar
1986	Paul John Flory
1985	William Draper Harkins
1984	William Bryan Weigand

1983	Giulio Natta
1982	George Oenslager
1981	Joseph C. Patrick
1980	Graham Moore
1979	Walter Bock
1978	No award given
1977	Misazo Tamamoto
1975–76	Arthur Victor Tobolsky
1974	John Boyd Dunlop and Robert William Thomson
1973	Charles Dufraisse
1972	George Stafford Whitby
1971	Melvin Mooney
1970	Kurt Otto Hans Meyer and Wemer Kuhn
1969	Ernest Harold Farmer
1967–68	Peter J. W. Debye
1966	Hermann Staudinger
1965	Johan Rudolph Katz
1964	Sir William A. Tilden
1963	Wallace Hume Carothers
1962	Guiseppe Bruni
1961	Carl Dietrich Harries
1960	Henry Nicholas Ridley
1959	Thomas Hancock
1958	C. Greville Williams, Carl O. Weber, Ian I. Ostromislensky, Charles Goodyear, and Henri Bouasse

APPENDIX 4

Rubber Division Best Paper Awards, 1955–2007

In 1955, the Rubber Division created the Best Paper Committee with the objective of improving the quality of technical presentations by evaluating and publicly recognizing outstanding presentations. The Best Paper Award recognizes those individuals authoring and presenting papers at Rubber Division meetings. Each award consists of a plaque and recognition at the Business and Awards meeting.

Selections for these awards are made by the Best Paper Committee on the recommendation of a number of peer judges who review the individual papers and attend the technical sessions. Clarity of presentation and significance of technical content are of prime importance to the judges. The papers are also judged on the adherence to the allotted time, SI units, and emphasis on the technical rather than promotional aspects of the paper.

MEETING	AWARD	AUTHORS	AFFILIATION	PAPER TITLE
Spring 1955	Best Paper	J. S. Rugg and G. W. Scott	DuPont	Adipene B, Urethane II: Factors Influencing Its Processability
Fall 1955	Best Paper	G. K. Sutherland and J. P. McKenzie	University of Michigan	A Glass Polymerization Vessel for Small Scale Laboratory Studies
Spring 1956	Best Paper	H. G. Dawson	Firestone	A Zinc Oxide Test for Hevea Latex
Fall 1956	Best Paper	S. W. Boggs and W. P. Rieman	U.S. Rubber	Physics on the Friction of Rubber on Rough Surfaces
Spring 1957	Best Paper	W. G. Forbes and L. A. McLeod	Polymer	Dependence on Tack Strength on Molecular Properties

231

MEETING	AWARD	AUTHORS	AFFILIATION	PAPER TITLE
Fall 1957	Best Paper	H. E. Diem, H. Tucker, and C. F. Gibbs	B. F. Goodrich	Cis-1, 4 Polyisoprene Rubber by Alkyl Lithium Polymerization
Spring 1958	Best Paper	G. Krauss and R. L. Collins	Phillips Petroleum	Odd Electrons in Rubber Reinforcing Blacks
Fall 1958	Best Paper	R. A. Pike, T. C. Williams, and F. A. Fekete	Union Carbide	Cyanosilicone Elastomers–A New Class of Solvent Resistant, High Temperature Rubbers
Spring 1959	Best Paper	H. W. Kilbourne, G. R. Wilder, J. E. Van Verth, and J. O. Harris	Monsanto	Chemical Inhibition of Ozone Degradation of SBR
Spring 1960	Best Paper	W. A. Smith and J. M. Willis	Firestone	Diene Rubber-Compounding and Testing
Fall 1960	Best Paper	R. M. Murray and J. D. Detenber	DuPont	A Study for First- and Second-Order Transitions in Neoprene
Spring 1961	Best Paper	W. M. Hess	Columbian Carbon	The Analysis of Pigment Dispersion in Rubber by Means of Light Microscopy, Microradiography, and Electron Microscopy
Fall 1961	Best Paper	E. K. Gladding, B. S. Fisher, and J. W. Collette	DuPont	A New Hydrocarbon Elastomer-I
Spring 1962	Best Paper	P. E. Wei and J. Rehner Jr.	Esso	New Vulcanizing Agents for Ethylene-Propylene Elastomers-II
Fall 1962	Best Paper	W. M. Hess and K. A. Burgess	Columbian Carbon	Groove Cracking in Tire Treads–A Microscopic Study
Spring 1963	Best Paper	W. F. Brucksch	U.S. Rubber	Vinyl Pyridine Rubber (PBR)–Filler Interactions
Fall 1963	Best Paper	W. E. Claxton and J. W. Liska	Firestone	Calculations of State of Cure in Rubber Under Variable Time-Temperature Conditions
Spring 1964	Best Paper	J. E. Callan, W. E. Ford, and B. Topak	Columbian Carbon	Butyl Blends–Processing and Properties
Fall 1964	Best Paper	B. M. Vanderbilt, and R. E. Clayton	Enjay	The Bonding of Fibrous Glass to Elastomers
Fall 1964	Honorable Mention	M. Gippin	Firestone	Stereoregular Polymerization of Butadiene with RAICl3 and Cobalt Octoate
Fall 1964	Honorable Mention	S. A. Banks, F. Brzenk, J. A. Rae, and C. S. Hwa	Enjay	Effect of Intracarcass Pressure Buildup on Tubeless Tire Performance
Fall 1964	Honorable Mention	W. W. Barbin	Firestone	Vulcanization Characteristics of Polybutadiene
Fall 1964	Honorable Mention	M. A. Dudley and A. J. Wallace	Enjay	Compounding Chlorobutyl Rubber for Heat Resistance
Spring 1965	Best Paper	L. W. Gamble, L. Westerman, and E. A. Knipp	Esso	Molecular Weight Distribution of Elastomers: Comparison of GPC with Other Techniques

MEETING	AWARD	AUTHORS	AFFILIATION	PAPER TITLE
Spring 1965	Honorable Mention	R. J. Athey	DuPont	Compounding Liquid Urethane Elastomers for Compression Molding
Spring 1965	Honorable Mention	W. M. Hess, P. A. Marsh, and F. J. Eckert	Columbian Carbon	Carbon-Elastomer Adhesion: (1) Analysis, (2) Influencing Factors
Spring 1965	Honorable Mention	F. A. Heckmann and D. F. Darling	Cabot	Progressive Oxidation of Selected Particles of Carbon Black: Further Evidence for a New Microstructural Model
Spring 1965	Honorable Mention	R. R. Barnhart and J. C. Mitchell	U.S. Rubber	New Accelerators for EPDM Compounds
Fall 1965	Best Paper	E. M. Dannenberg and J. J. Brennan	Cabot	Strain Energy as Criterion for Stress Softening in Carbon Black Filled Vulcanizates
Fall 1965	Honorable Mention	H. M. Cole, D. L. Petterson, V. A. Sljaka, and P. S. Smith	Cabot	Identification of Polymers in Compounded Cured Rubber Stock by Pyrolysis/Two Channel Gas Chromatography
Fall 1965	Honorable Mention	J. D. Skewis	U.S. Rubber	Self-Diffusion Coefficients and Tack Properties of Some Rubbery Polymers
Fall 1965	Honorable Mention	B. A. Hunter and M. J. Kleinfeld	U.S. Rubber	Low Temperature Expansion of Polysulfide Rubbers
Fall 1965	Honorable Mention	C. Booth	Esso	Factors Influencing Rubber Injection Molding Process
Spring 1966	Best Paper	W. R. Griffin	Wright Patterson AFB	Triazine Elastomers
Spring 1966	Honorable Mention	C. L. Whipple and J. A. Thorne	Rubbermaid	Performance of Elastomeric Silicone in Ablative and Space Environments
Spring 1966	Honorable Mention	R. I. Leininger, R. D. Falb, and G. A. Grode	Battelle	Performance of Elastomers in the Human Body
Spring 1966	Honorable Mention	E. M. Bevilacqua	Uniroyal	Aging of SBR II
Spring 1966	Honorable Mention	S. D. Gehman and G. M. Larsen	Goodyear	Rapid Treadwear Ratings with Radioactive Isotopes
Spring 1966	Honorable Mention	G. R. Cotten	Cabot	Carbon Black Reinforcement in Preswollen Rubbers
Fall 1966	Best Paper	W. M. Hess, C. E. Scott, and J. E. Callan	Columbian Carbon	Carbon Black Distribution in Elastomer Blends
Fall 1966	Honorable Mention	H. K. J. de Decker and K. J. Sabatini	Texas-U.S.	Dynamic Properties of Elastomer Blends
Fall 1966	Honorable Mention	G. W. Ross, J. F. Svetlik, D. D. Dearmont, and M. H. Richmond	Phillips	Solution Butadiene-Styrene Block Compounds in Blends for Specific Applications
Fall 1966	Honorable Mention	J. M. Willis, and R. L. Denecour	Firestone	Tire Tread Application of EPDM-Butyl Blends

MEETING	AWARD	AUTHORS	AFFILIATION	PAPER TITLE
Fall 1966	Honorable Mention	C. W. Snow and F. W. Barlow	United Carbon	Curing Rates of Carbon Blacks in SBR by Classical and Modern Methods
Spring 1967	Best Paper	V. F. Fischer and W. H. King Jr.	Enjay	New Method of Measuring Oxidation Stability of Elastomers
Spring 1967	Honorable Mention	W. M. Hess, L. L. Ban, F. J. Eckert, and V. Chirico	Columbian Carbon	Microstructural Variations in Commercial Carbon Blacks
Spring 1967	Honorable Mention	R. A. Livigni, L. Marker, G. Shkapenko, and S. L. Aggarwal	General Tire	Structure and Transition Behavior of Isoprene-Styrene Copolymers of Different Sequence Length
Spring 1967	Honorable Mention	V. D. Mochel	Firestone	NMR Compositional Analysis of Copolymers
Spring 1967	Honorable Mention	M. Morton, P. C. Juliano, J. E. McGrath	University of Akron	Structure-Property Relations of Styrene-Diene Thermoplastic Elastomers
Spring 1967	Honorable Mention	G. Vitolins	Enjay	Adhesion of Chlorobutyl Blends to Highly Unsaturated Elastomers. Effects of Vulcanization and Carbon Blacks
Fall 1967	Best Paper	M. L. Deviney and L. E. Whittington	Ashland	Radiotracer Studies of Carbon Black Surface Interactions with Organic Systems
Fall 1967	Honorable Mention	J. W. Horvath, W. A. Wilson, H. S. Lundstrom, and J. R. Purdon	Goodyear	Investigation of Factors Influencing Properties of NBR/PVC Blends
Fall 1967	Honorable Mention	P. A. Marsh, A. Voet, and L. D. Price	Columbian Carbon, Center Research CNRS	Electron Microscopy of Heterogeneous Elastomer Blends
Fall 1967	Honorable Mention	R. A. Buchanan, O. E. Weislogel, C. R. Russell, and C. E. Rist	Northern Research Lab., U.S. Dept. of Agriculture, Peoria, Illinois	Starch in Rubber. I. Zinc Starch Xanthate in Later Masterbatching
Fall 1967	Honorable Mention	R. D. Stiehler, E. J. Parks, and F. J. Linnig	National Bureau of Standards	Stiffening of Elastomers by Organic Fillers
Fall 1967	Honorable Mention	B. L. Williams and L. Weissbein	American Cyanamid	Torsional Braid Analysis of Antioxidant Activity in Elastomer Systems
Spring 1968	Best Paper	J. B. Gardiner	Enjay	Curative Diffusion Between Dissimilar Elastomers and Its Influence on Adhesion
Spring 1968	Honorable Mention	E. P. Percarpio and E. M. Bevilacqua	Uniroyal	Fundamental Aspects of Lubricated Friction of Rubber, Part VI
Spring 1968	Honorable Mention	M. L. Deviney, J. E. Lewis, and B. G. Corman	Ashland, Esso	Migration of Extender Oil in Natural and Synthetic Rubber. III.
Fall 1968	Best Paper	G. Kraus and J. T. Gruver	Phillips	Viscosity of Carboxy-Terminated Polybutadienes

MEETING	AWARD	AUTHORS	AFFILIATION	PAPER TITLE
Fall 1968	Honorable Mention	W. H. Whittington, M. L. Deviney, and J. E. Lewis	Ashland	Migration of Antioxidants and Accelerators in Natural and Synthetic Rubber. II. Sulfenamide System Studies and Comparison of Age Resistor Migration Under Inert and Practical Conditions
Fall 1968	Honorable Mention	W. F. Fischer	Enjay	Profile Vulcanization–New Method for Measuring Total Vulcanization History of Elastomers
Spring 1969	Best Paper	J. C. Ambelang, F. H. Wilson, L. E. Porter, and D. L. Turk	Goodyear	Ozone Crack Initiation–Statistical and Deterministic Interpretation of Action of p-Phenylenediamines and EPDM as Antiozonants
Spring 1969	Honorable Mention	R. W. Wise and W. R. Deason	Monsanto	Programmed Temperature Curemetry
Spring 1969	Honorable Mention	I. Prettyman and M. J. Forster	Firestone	New Improved Tire Cord Fatigue Tester
Fall 1969	Best Paper	C. S. L. Baker, O. Barnard, and M. Porter	NRPRA	New Reactions for Vulcanization of Natural Rubber
Fall 1969	Honorable Mention	W. F. Fischer	Enjay	Profile Vulcanization–II, Inconstant Nature of Elastomer Vulcanization
Fall 1969	Honorable Mention	F. P. Baldwin, D. Borzel, C. A. Cohen, M. S. Makowski, and J. F. Van DeCastle	Enjay, Esso	Influence of Residual Olefin Structure on EPDM Vulcanization
Spring 1970	Best Paper	F. K. Haas	Bayer	Properties of a 1,5-trans-Polypentenamer Produced by Polymerization through Ring Cleavage Cyclopentene
Spring 1970	Honorable Mention	F. J. Kovac	Goodyear	Technological Forecasting
Spring 1970	Honorable Mention	R. I. Leib, A. B. Sullivan, and C. D. Trivette Jr.	Monsanto	Prevulcanization Inhibitor: The Chemistry of Scorch Delay
Fall 1970	Best Paper	J. E. Lewis, H. M. Mercer, M. Deviney, L. Hughes, and J. F. Jewell	Ashland	Autoradiographic Electron Microscopy of Elastomer Blends
Fall 1970	Honorable Mention	J. L. Koenig, J. R. Shelton, and M. M. Coleman	Case Western Reserve University	Raman Spectra of Vulcanized Rubber
Fall 1970	Honorable Mention	V. L. Folt	B. F. Goodrich	Crystallization of cis-Polyisoprenes in a Capillary Rheometer
Spring 1971	Best Paper	R. T. Morrissey	B. F. Goodrich	Halogenation of EPDM
Spring 1971	Honorable Mention	J. R. Shelton, J. L. Koenig, and M. M. Coleman	Case Western Reserve University	Raman Studies
Spring 1971	Honorable Mention	M. L. Janssen and J. D. Walter	Firestone	Strain Measurements on Tires

MEETING	AWARD	AUTHORS	AFFILIATION	PAPER TITLE
Fall 1971	Best Paper	N. G. Bartfug	PPG Industries	High DPU–Major Factor in Good Fiberglass Tire Cord Performance
Fall 1971	Honorable Mention	M. C. Throckmorton and F. S. Farson	Goodyear	An HF-Ni-R3Al Catalyst System for Producing High-cis 1,4-Polybutadienes
Fall 1971	Honorable Mention	K. T. Potts and E. G. Bruge	Rensselaer Polytechnic Institute	Electron Impact Induced Pigmentations of Benzothiazole Sulfenamides and Related Compounds: Correlation with Rubber Accelerator Activity. Determination of Head-Head and Tail-Tail Structures in Polyisoprene
Spring 1972	Best Paper	M. J. Brock and M. J. Hackathorn	Firestone	Synergistic Effect of NR on Crystallization of Lithium Polyisoprene
Spring 1972	Honorable Mention	A. I. Medalia	Cabot	Effective Degree of Immobilization of Rubber Occluded within Carbon Black Aggregates
Spring 1972	Honorable Mention	F. W. Wilcox	Witco	Millable Urethane Elastomers
Fall 1972	Best Paper	R. L. Zapp	Exxon	Chlorobutyl Covulcanization Chemistry and Interfacial Bonding
Fall 1972	Honorable Mention	R. F. Hoffman and R. H. Gobpan	Thiokol	A New Role for Liquid Carboxyl Terminated Polybutadiene Polymers
Fall 1972	Honorable Mention	G. E. Meyer, R. W. Kauchok, and F. J. Naples	Goodyear	Emulsion Rubbers with Copolymerized Monomeric Antioxidants
Spring 1973	Best Paper	J. C. Falk, R. J. Schlott, and D. F. Hoeg, and J. F. Pendleton	Borg-Warner	Thermoplastic Elastomer-Styrene Grafts on Lithiated Polydiene plus Hydro-generated Counterparts
Spring 1973	Honorable Mention	T. C. Bouton and S. Futamura	Firestone	Solution SBR-Study in Copolymerization Dynamics
Spring 1973	Honorable Mention	A. C. Patel and M. Deviney	Ashland	Precision Treadwear Measurements at Increased Mileages Via Improved Radioiodine Method
Fall 1973	Best Paper	K. C. Baranwal and P. N. Son	B. F. Goodrich	Cocuring Blends of EPDM plus Diene Rubbers by Grafting Accelerators onto EPDM
Fall 1973	Honorable Mention	R. W. Seymour and S. L. Cooper	University of Wisconsin	Viscoelastic Properties of Polyurethane Block Polymers
Fall 1973	Honorable Mention	F. K. Lautenschlaeger	Dunlop	Observations on Crosslinking of NR with Nitrosophenols plus Diisocyanates
Spring 1974	Best Paper	J. R. Dunn	Polysar	Review of Unsolved Problems in Protection of Rubber against Oxidative Degradation
Spring 1974	Honorable Mention	H. E. Railsback and N. A. Stumpe	Phillips	Blends of Medium Vinyl Polybutadienes with SBR Rubbers
Spring 1974	Honorable Mention	J. R. Shelton	Case Western Reserve University	Role of Certain Organic Sulfur Compounds as Preventive Antioxidants

MEETING	AWARD	AUTHORS	AFFILIATION	PAPER TITLE
Fall 1974	Best Paper	D. W. Brazier and G. H. Nickel	Dunlop	Thermoanalytical Methods in Vulcanizate Analysis I: Differential Scanning Calorimetry and Heat of Sulfur Vulcanization
Fall 1974	Honorable Mention	H. A. Pfisterer and J. R. Dunn	Polysar	Assessing the Performance of Oil-Resistant Vulcanizates at High Temperatures
Fall 1974	Honorable Mention	J. C. Andries, D. B. Ross, and H. E. Deim	B. F. Goodrich	Ozone Attack and Anitozonant Protection of Vulcanized Natural Rubber Compounds: A Surface Study by Attenuated Total Reflectance Spectroscopy
Spring 1975	Best Paper	N. M. van Gulick, J. J. Verbanc, P. P. Caruso, and R. J. Athey	DuPont	Coordination Complexes of Methylene Dianiline-Chemistry and Applications
Spring 1975	Honorable Mention	H. Cohen, J. A. Rae, and D. J. Buckley	Exxon	High Molecular Weight, Highly Unsaturated Isobutylene-Conjugated Diene Copolymers. A New Family of Elastomers. II. Properties and Performance
Spring 1975	Honorable Mention	M. Morton and D. C. Rubio	University of Akron	Synthesis and Properties of Uniform Polyisoprene Networks
Fall 1975	Best Paper	R. T. Morrissey	B. F. Goodrich	Sulfur-Cure Compatible Blends of Halogenated Ethylene-Propylene Copolymers and Diene Rubber
Fall 1975	Honorable Mention	D. F. Brizzolara, W. Honsberg, and R. W. Keown	DuPont	A New Carboxylated Neoprene Latex
Fall 1975	Honorable Mention	M. Kato and J. E. Mark	University of Michigan	The Effects of Temperature on the Stress-Strain Isotherms of Polyisobutylene Networks at High Elongation
Spring 1976	Best Paper	S. L. Cooper, J. C. West, and A. Lilaonitkul	University of Wisconsin	Properties of Polyether-Polyester Thermoplastic Elastomers
Spring 1976	Honorable Mention	J. G. Sommer, H. N. Grover, and P. T. Suman	General	In-place Cleaning of Rubber Curing Molds
Spring 1976	Honorable Mention	R. M. Pierson, T. G. Gurley, D. V. Hillegass, and R. J. Arconti	Goodyear	Research on Uses of Rubber in the Cardiovascular System
Fall 1976	Best Paper	B. B. Boonstra	Cabot	Resistivity of Unvulcanized Rubber Carbon Black Compounds
Fall 1976	Honorable Mention	M. P. Wagner	PPG Industries	Silica—A Petroleum-Free Reinforcing Filler
Fall 1976	Honorable Mention	R. A. Buchanan, I. M. Cull, F. H Otey, and C. R. Russell	U.S. Dept. of Agriculture	Hydrocarbon and Rubber Producing Crops: Evaluation of U.S. Plant Species
Spring 1977	Best Paper	G. C. McDonald and W. M. Hess	Columbian Chemicals	Carbon Black Morphology in Rubber

MEETING	AWARD	AUTHORS	AFFILIATION	PAPER TITLE
Spring 1977	Honorable Mention	J. F. Krymouski and R. D. Taylor	B. F. Goodrich	Chemical Reactions between Thiocarbamylsulfenamides and Benzothiazolesulfenamides
Spring 1977	Honorable Mention	J. D. Byam and G. P. Colbert	DuPont	Applying Science to Processing Profitably. Part I: Injection Molding
Fall 1977	Best Paper	J. R. Dunn, D. C. Coulthard, and H. A. Pfisterer	Polysar	Advances in Nitrile Rubber Technology: A Review
Fall 1977	Honorable Mention	P. A. Lindfors, W. M. Riggs, and L. E. Davis	Physical Electronics Industries	Application of AES, XPS, and SIMS to the Study of Adhesion Related Metal and Polymer Surface Composition
Fall 1977	Honorable Mention	R. R. Rahalkar, C. U. Yu, and J. E. Mark	University of Michigan	The Ultimate Properties of Unswollen Polydimethylsiloxane Networks and Their Dependence on the Degree of Crosslinking and the Amount of Diluent Present during the Crosslinking Process
Spring 1978	Best Paper	E. J. Buckler, G. J. Briggs, J. R. Dunn, E. Lasis, and Y. K. Weis	Polysar	Green Strength in Emulsion SBR
Spring 1978	Honorable Mention	P. C. Vegvari, W. M. Hess, and V. E. Chirco	Columbian Chemicals	Measurement of Carbon Black Dispersion in Rubber by Surface Analysis
Spring 1978	Honorable Mention	F. K. Lautenschlaeger and P. Zeeman	Polysar	Model Compound Vulcanization, Part II: Accelerator Comparisons for Natural Rubber
Fall 1978	Best Paper	J. R. Falender, G. S. Y. Yeh, J. E. Mark	University of Michigan, University of Cincinnati	The Effect of Crosslink Distribution on Elastomeric Properties
Fall 1978	Honorable Mention	D. Barnard	Malaysian Rubber Producers Association	Natural Rubber Research Perspectives and Strategies
Fall 1978	Honorable Mention	J. R. Dunn, H. A. Pfisterer, and J. J. Ridland	Polysar	NBR Vulcanizates Resistant to High Temperature and Peroxidized Gasoline
Spring 1979	Best Paper	D. M. Coddington	Exxon Chemical	Inflation Pressure Loss in Tubeless Tires–Effects of Tire Size, Service and Construction
Spring 1979	Honorable Mention	L. A. Goettler, R. I. Leib, and A. J. Lambright	Monsanto	Short Fiber Reinforced Hose–A New Concept in Production and Performance
Spring 1979	Honorable Mention	N. L Hewitt and M. P. Wagner	PPG Industries	New Developments in a Dynamic Wire Adhesion Test
Fall 1979	Best Paper	R. P. Lattimer and K. R. Welch	B. F. Goodrich	Direct Analysis of Polymer Chemical Mixtures by Field Desorption Mass Spectroscopy
Fall 1979	Honorable Mention	D. C. Edwards and K. Sato	Polysar	The Interaction of Silica with Functionalized SBR

MEETING	AWARD	AUTHORS	AFFILIATION	PAPER TITLE
Fall 1979	Honorable Mention	J. Janzen and G. Kraus	Phillips Petroleum	New Methods for Estimating Dispersibility of Carbon Blacks in Rubber
Spring 1980	Best Paper	D. D. Werstler	General	Analysis of Cured, Filled Elastomeric Compounds by 13C NMR
Spring 1980	Honorable Mention	R. P. Lattimer, E. R. Hooser, H. E. Diem, R. W. Layer, and C. K. Rhee	B. F. Goodrich	Mechanisms of Ozonation of N,N'-di-(1-methylheptyl)-p-phenylenediamine
Spring 1980	Honorable Mention	C. K. Rhee and J. C. Andries	B. F. Goodrich	Factors Which Influence Autohesion of Elastomers
Fall 1980	Best Paper	G. R. Hamed	University of Akron	Tack and Green Strength of NR, SBR, and NR/SBR Blends
Fall 1980	Honorable Mention	L. A. Goettler, R. I. Leib, P. J. DiMauro, and A. J. Lambright	Monsanto	Extrusion-Shaping of Curved Hose Reinforced with Short Cellulose Fibers
Fall 1980	Honorable Mention	C. R. Wilder, J. R. Haws, and W. T. Cooper	Phillips Petroleum	Effects of Carbon Black Types on Treadwear of Radial and Bias Tires at Variable Test Severities
Spring 1981	Best Paper	T. H. Kuan and J. G. Sommer	General	Water-resistant Release Coating for Uncured Rubber
Spring 1981	Honorable Mention	M. T. Maxson and C. L. Lee	Dow Corning	Effects of Fumed Silica Treated with Functional Disilazanes on Silicone Elastomer Properties
Spring 1981	Honorable Mention	R. S. Ro	DuPont	Vinyl Adhesives from Neoprene/Methyl Methacrylate Graft Polymers
Fall 1981	Best Paper	R. C. Hirst	Goodyear	Applications of Computerized Infrared and Nuclear Magnetic Resonance Spectroscopic Techniques to Rubber Analysis
Fall 1981	Honorable Mention	J. J. Maurer and D. W. Brazier	Exxon Research, Dunlop Research	Applications of Thermal Analysis in the Rubber Industry
Fall 1981	Honorable Mention	R. W. Smith and A. G. Veith	B. F. Goodrich	Electron Microscopical Examination of Worn Tire Treads and Tread Debris
Spring 1982	Best Paper	H. L. Stephens and T. H. Hashmi	University of Akron	Prevulcanization of Natural Rubber Latex in Dipped Goods Manufacture
Spring 1982	Honorable Mention	L. Spenadel	Exxon Chemical	Heat Aging Performance of Ethylene Propylene Elastomers in Electrical Insulation Compounds
Fall 1982	Best Paper	G. R. Hamed and T. Donatelli	University of Akron	Effect of Accelerator Type on Brass-Rubber Adhesion
Fall 1982	Honorable Mention	M. E. Martini	Firestone	Passenger Tire Rolling Loss: A Tread Compounding Approach and Its Tradeoffs
Fall 1982	Honorable Mention	G. L. Finley and H. J. Herzlich	Armstrong	Tredloc-A Unique Engineering Technology for Radial Ply Tires
Spring 1983	Best Paper	A. N. Gent	University of Akron	Fracture Mechanics Applied to Elastomeric Composites

MEETING	AWARD	AUTHORS	AFFILIATION	PAPER TITLE
Spring 1983	Honorable Mention	L. E. Porter	Goodyear	The Isolation and Identification of Contaminants in Rubber by Microscopical Techniques
Spring 1983	Honorable Mention	W. E. Thibodeau, P. S. Johnson, and K. R. Gark	Polysar Limited	An Application of Microcomputers in Rubber Mixing
Fall 1983	Best Paper	W. M. Hess, R. A. Swor, and E. J. Micek	Columbian Chemicals	The Influence of Carbon Black, Mixing and Compounding Variables on Dispersion
Fall 1983	Honorable Mention	J. D. Skewis	Uniroyal	The Influence of Environmental Conditions on Tack of Rubber Compounds
Fall 1983	Honorable Mention	G. R. Hamed and C.-H. Shieh	University of Akron	Tack and Related Properties of Isopropyl Azodicarboxylate Modified Polybutadiene
Spring 1984	Best Paper	G. Anthoine, E. R. Lynch, D. E. Mauer, and P. G. Moniotte	Monsanto	A New Concept to Stabilize Cured NR Properties During Thermal Aging and Improve Adhesion to Brass
Spring 1984	Honorable Mention	C. S. L. Baker, I. R. Gelling, and R. L. Newell	MRPRA	Epoxidized Natural Rubber
Spring 1984	Honorable Mention	H. L. Hsieh and H. C. Yeh	Phillips Petroleum	Butadiene and Isoprene Polymers and Copolymers Prepared with Lanthanide Catalysts
Fall 1984	Best Paper	R. J. Eldred	General Motors	Plasticization by In Situ Grafted Acrylates II
Fall 1984	Honorable Mention	R. G. Stacer, L. C. Yanyo, and F. N. Kelley	University of Akron	Observations on the Tearing of Elastomers
Fall 1984	Honorable Mention	R. Vukov and G. J. Wilson	NACAN Products, Polysar	Crosslinking Efficiencies of Some Halobutyl Curing Reactions
Spring 1985	Best Paper	S. Bagrodia, G. L. Wilkes, and J. P. Kennedy	Virginia State University, University of Akron	Solid State Mechanical Properties of Sulfonated Telechelic Polyisobutylene-Based Elastomeric Ionomers
Spring 1985	Honorable Mention	D. Y. Sogah, O. W. Webster, and W. R. Hertler	DuPont	Synthesis of Block Copolymers by Group Transfer Polymerization
Spring 1985	Honorable Mention	W. P. Gergen, S. Davidson, and R. G. Lutz	Shell	Thermoplastic Elastomer IPNs
Fall 1985	Best Paper	J. F. Stevenson	GenCorp	Multidimensional Control with a Single Roller Die
Fall 1985	Honorable Mention	C. B. Shulman	Exxon	Unique Features of High Unsaturation EPDM Polymers
Fall 1985	Honorable Mention	R. P. Lattimer, J. Gianelos, H. E. Diem, R. W. Layer, and C. K. Rhee	B. F. Goodrich	Mechanisms of Antiozonant Protection: Unextractable Nitrogen in Aged cis-Polybutadiene Vulcanizates
Spring 1986	Best Paper	D. G. Young	Exxon	Fatigue Crack Propagation in Elastomer Compounds: Effects of Strain Rate, Temperature, Strain Level, and Oxidation

MEETING	AWARD	AUTHORS	AFFILIATION	PAPER TITLE
Spring 1986	Honorable Mention	R. P. Lattimer, R. E. Harris, C. K. Rhee	B. F. Goodrich	Identification of Organic Additives in Rubber Vulcanizates by Using Mass Spectrometry
Spring 1986	Honorable Mention	D. C. Edwards	Polysar, Ltd.	A High-Performance Curing System for Halobutyl Elastomers
Fall 1986	Best Paper	R. W. Layer	B. F. Goodrich	A Postcrosslinking Accelerator System for Natural Rubber Based on Thiocarbamyl Sulfenamides
Fall 1986	Honorable Mention	L. C. Yanyo and F. N. Kelley	Lord, University of Akron	Effect of Chain Length Distribution of the Tearing Energy of Silicone Elastomers
Fall 1986	Honorable Mention	D. Chaembaere and W. Coppens	NV Bekaert S. A.	Determination and Influences of the Electrical Resistances Involved in the Corrosion of the Rubber-Steel Cord Composite
Spring 1987	Best Paper	T. W. Besuden and L. L. Beumel	Naval Research, Texas Research	Resistivity of Rubber as a Function of Mold Pressure
Spring 1987	Honorable Mention	L. G. Gatti	Armstrong	Acceleration Dependent Hysteresis, Nonvibration Testing for Viscoelastic Properties
Spring 1987	Honorable Mention	A. M. Zaper and J. L. Koenig	Case Western Reserve University	Solid State Carbon-13 NMR Studies of Accelerated Sulfur-Vulcanized Natural Rubber
Fall 1987	Best Paper	M. D. Ellul and R. J. Emerson	GenCorp	A New Pull-Out Test for Tire Cord Adhesion-Part I. Hot Bonding
Fall 1987	Honorable Mention	M. D. Ellul and R. J. Emerson	GenCorp	A New Pull-Out Test for Tire Cord Adhesion-Part II Cold Bonding
Fall 1987	Honorable Mention	R. J. Eldred	General Motors	Effect of Oriented Platy Filler on the Fracture Mechanism of Elastomers
Spring 1988	Best Paper	R. B. Layer	B. F. Goodrich	Synergism between Thiocarbamyl Sulfenamide and 2-Benzothiazyl Sulfenamide Accelerators in Various Rubbers
Spring 1988	Best Paper —Review	R. P. Lattimer and R. E. Harris	B. F. Goodrich	Analysis of Components in Rubber Compounds Using Mass Spectrometry
Spring 1988	Honorable Mention —Original Contribution	R. P. Lattimer, R. A. Kinsey, and R. W. Layer	B. F. Goodrich	The Mechanism of Phenolic Resin Vulcanization of Unsaturated Elastomers
Spring 1988	Honorable Mention	M. Gerspacher and C. M. Lansinger	Goodyear	Carbon Black Characterization and Application to Study Polymer/Filler Interaction
Fall 1988	Best Paper —Original Contribution	K. L. Rollick, J. G. Gillick, J. L. Bush, and J. A. Kuczkowski	Goodyear	Triazine Thiones: A New Class of Nonstaining Nondiscoloring Antiozonants
Fall 1988	Best Paper —Review	K. Cas and J. S. Bigelow	Exxon Exxon	Capability and Performance Indices: Proper Use in the Process Industries

MEETING	AWARD	AUTHORS	AFFILIATION	PAPER TITLE
Fall 1988	Honorable Mention –Original Contribution	A. J. Tinker and M. J. R. Loadman	Malaysian Rubber Producers	Estimation of the Extent of Crosslinking in the Phases of Vulcanized Rubber Blends By CW 1H NMR Spectroscopy
Fall 1988	Honorable Mention –Review	J. R. Dunn	Polysar, Ltd.	Compounding for Modern Automotive Requirements
Spring 1989	Best Paper –Original Contribution	J. V. Fusco, I. J. Gardner, P. Hons, and D. S. Tracey	Exxon	Chlorobutyl Designed for Curing Members
Spring 1989	Best Paper –Review	D. L. Hertz Jr.	University of Akron	Solubility Parameter Concepts
Spring 1989	Honorable Mention –Original Contribution	F. L. Magnus and G. R. Hamed	University of Akron	Role of Phenolic Tackifiers in Polyisoprene Rubber
Spring 1989	Honorable Mention –Review	D. I. Livingston	Goodyear	Factors Affecting Tire Mileage in Even Wear
Fall 1989	Best Paper –Original Contribution	D. K. Parker, H. A. Colvin, A. W. Weinstein and S. Chen	Goodyear	Reactively Curable Rubbers–I: Diene Elastomers with Pendant Isocyanate and/or Hydroxyl Functionality
Fall 1989	Best Paper –Review	C. S. L. Baker	MRPRA	Advances in Natural Rubber Science and Technology
Fall 1989	Honorable Mention –Original Contribution	W. M. Grootaert, R. E. Kolb, and A. T. Worm	3M	A Novel Fluorocarbon Elastomer for High-Temperature Sealing Applications in Aggressive Motor-Oil Environments
Spring 1990	Best Paper –Original Contribution	F. K. Jones, L. L. Outzs, and G. Liolios	DuPont	The Effect of Heat Generation on the Performance of Several Elastomers for Automotive Use
Spring 1990	Best Paper –Review	R. P. Quirk	University of Akron	Recent Advances in Anionic Synthesis of Functionalized Polymers
Spring 1990	Honorable Mention –Original Contribution	E. A. Meinecke	University of Akron	Effect of Crosslink Density and Carbon-Black Loading on the Heat Build-up in Elastomers
Fall 1990	Best Paper –Original Contribution	R. A. Swor, W. M. Hess, and E. J. Micek	Columbian Chemical	New High-Structure Tread Black for Compounding Flexibility
Fall 1990	Best Paper –Review	R. P. Quirk and J. Kim	University of Akron	Recent Advances in Thermoplastic Elastomer Synthesis
Fall 1990	Honorable Mention –Original Contribution	R. D. Vargo and F. N. Kelley	University of Akron	Properties of Highly Filled, Highly Plasticized, Semicrystalline Polymer Networks

MEETING	AWARD	AUTHORS	AFFILIATION	PAPER TITLE
Spring 1991	Best Paper —Original Contribution	A. Y. Coran and S. Lee	Monsanto	New Elastomers by Reactive Processing. Part II Dynamic Vulcanization of Blends by Trans-Esterification
Spring 1991	Best Paper —Review	F. L. McMillian	DuPont	Neoprene Latexes and their Applications
Spring 1991	Honorable Mention —Original Contribution	D. K. Parker, H. W. Schiessl, and R. F. Roberts	Goodyear	A New Process for the Preparation of Highly Saturated Nitrile Rubber in Latex Form
Fall 1991	Best Paper —Original Contribution	R. C. Keller and D. A. White	Exxon	Electrochemical Influence on Carbon-Filled Ethylene-Propylene Elastomers
Fall 1991	Best Paper —Review	A. Limper and W. Haeder	Werner and Pfleiderer Gummitechnix	The Mixing Room under System Aspects
Fall 1991	Honorable Mention —Original Contribution	R. W. Layer	B. F. Goodrich	Recuring Vulcanizates II. Effect of Added Accelerators
Fall 1991	Best Symposium	H. Young	Copolymer Rubber and Chemical	Polymer-Fluid-Filler Interactions
Spring 1992	Best Paper —Original Contribution	J. A. Sezna	Monsanto	Rubber Testing for Injection Molding
Spring 1992	Best Paper —Review	A. D. Roberts	MRPEA	Winter and All-Season Tires
Spring 1992	Honorable Mention —Original Contribution	O. K. Gardner, J.-C. Wang, and J. V. Fusco	Exxon	Crosslinking of Brominated p-Methylstyrene-Isobutylene Copolymers in Blends with General Purpose Rubbers
Spring 1992	Honorable Mention —Review	D. G. Rowland	Uniroyal	Expanding Rubber Through the Years
Spring 1992	Best Symposium	R. J. Hupp and W. B. Lee	Huls America, Lloyd Manufacturing	Advances in the Injection Molding of Rubber
Fall 1992	Best Paper —Original Contribution	C. R. Herd, G. C. McDonald, R. E. Smith, and W. M. Hess	Columbian Chemicals	The Use of Skeletonization for the Shape Classification of Carbon Black Aggregates
Fall 1992	Best Paper —Review	R. Eller	Charles River Associates	Thermoplastic Elastomers in the Global Automotive Industry
Fall 1992	Honorable Mention —Original Contribution	A. Y. Coran and S. Lee	University of Akron, Monsanto	High Performance Elastomers Prepared by Dynamic Vulcanization
Fall 1992	Best Symposium	J. R. Dunn	Polysar	Elastomers for Automotive Applications

MEETING	AWARD	AUTHORS	AFFILIATION	PAPER TITLE
Spring 1993	Best Paper –Original Contribution	C. J. Rostek, H. J. Lin, and D. J. Sikora	Monsanto, University of Florida	Novel Sulfur Vulcanization Accelerators Based on Mercaptopyridines
Spring 1993	Best Paper –Review	J. Ryba and M. B. Rodgers	Goodyear	Materials Requirements for Special Service Tires in Mining Applications
Spring 1993	Honorable Mention –Original Contribution	H. G. Kilian, M. Strauss, and W. Hamm	University of Ulm	Universal Properties of Filler-Loaded Rubbers
Spring 1993	Best Symposium	T. G. Hutchins	Goodyear	Elastomers in Mining
Fall 1993	Best Paper –Original Contribution	D. K. Parker and R. F. Roberts	Goodyear	The Preparation, Properties and Potential Applications of Diimide-Hydrogenated Styrene-Butadiene (HSBR) and Polybutadiene (HBR) Thermoplastic Elastomers
Fall 1993	Best Paper –Review	J. R. Dunn	J. R. Dunn Consulting	Overview of Rubber Recycling
Fall 1993	Honorable Mention –Original Contribution	H. L. Hsu and A. F. Halasa	Goodyear	Preparation and Characterization of Crystalline 3,4-Polyisoprene
Fall 1993	Honorable Mention –Review	M. Beck	Polymer Valley Chemical	Pyrolysis Carbon-An Overview
Fall 1993	Best Symposium	C. Baker	MRPRA	Advances in Natural Rubber
Spring 1994	Best Paper –Original Contribution	G. J. Arsenault, T. A. Brown, and I. R. Jobe	Polysar Rubber	An Approach to Modern Polymer Development: Enhancement of the Service Temperature Range for Hydrogenated Nitrile-Butadiene Rubber (HNBR)
Spring 1994	Best Paper –Review	M. W. Rouse	Rouse Rubber Industries	Applications of Crumb Rubber Modifiers (CRM) in Asphaltic Materials
Spring 1994	Honorable Mention –Original Contribution	W. H. Waddell, L. R. Evans, and J. R. Parker	PPG Industries	Improved Farm Tread Performance Using Precipitated Silica
Spring 1994	Honorable Mention –Review	E. Fesus	Goldsmith and Eggleton	Scrap Rubber: A Compounding Tool
Spring 1994	Best Symposium	F. G. Smith	Environmental Technologies Alternatives	Rubber Recycling Topical Rubber
Fall 1994	Best Paper –Original Contribution	D. F. Lawson, K. J. Kim, and T. L. Fritz	Bridgestone/ Firestone	Chemical Modifications of Rubber Surfaces
Fall 1994	Best Paper –Review	S. B. Nablo and K. Makuuchi	Energy Science, Japan EERE	Techniques for Electron Vulcanization of Rubber

MEETING	AWARD	AUTHORS	AFFILIATION	PAPER TITLE
Fall 1994	Honorable Mention –Original Contribution	D. J. Sikora, H.-J. Lin, and O. Maender	Monsanto	Part I: New Ultra Accelerators-Vulcanization of High-Diene Rubbers
Fall 1994	Honorable Mention –Review	J. S. Dick	Monsanto	ASTM in the Globalization of Rubber Standards and Specifications
Fall 1994	Best Symposium	W. Waddell and M. Zaper	PPG Monsanto	Advanced Methodologies for Materials Testing
Spring 1995	Best Paper –Original Contribution	J. S. Dick and H. Pawlowski	Monsanto	Applications for the Curemeter Maximum Cure Rate in Rubber Compound Development and Process Control
Spring 1995	Best Paper –Review	W. H. Waddell and L. R. Evans	PPG	Review of Applications of Proton Induced X-Ray Emission Spectroscopy to Tire Material Characterizations
Spring 1995	Honorable Mention –Original Contribution	R. L. Warley	Cabot	Dynamic Properties of Elastomers as Related to Vibration Isolator Performance
Fall 1995	Best Paper –Original Contribution	W. H. Waddell, L. R. Evans, L. J. Snodgrass, and E. G. Goralski	PPG	Mechanism by which Precipitated Silica Improved Brass-Coated Wire-to-Natural Rubber Adhesion
Fall 1995	Best Paper –Review	A. I. Kasner and E. A. Meinecke	University of Akron	Porosity in Rubber: A Review
Fall 1995	Honorable Mention –Original Contribution	K. O. McElrath and M. B. Measmer	Exxon	The Role of Additives in Improving Compression Set of Cured Rubber Compounds
Fall 1995	Honorable Mention –Review	H. W. Gunn and M. Cole	H. L. Blachford, Ltd.	The Dollars and Sense of Anti-Track Coatings
Fall 1995	Best Symposium	F. Smith	ETA	International Review Program on Rubber Recycling
Spring 1996	Best Paper –Original Contribution	R. F. Ohm and T. C. Taylor	R. T. Vanderbilt, DuPont	An Improved Curing System for Chlorine-Containing Polymers
Spring 1996	Best Paper –Review	P. R. Dean	Harwick Chemical	Material Solution Options–The Total Package
Spring 1996	Honorable Mention –Original Contribution	A. S. Goeringer, W. A. Wampler, M. Gerspacher, and K Rajeshwar	Sid Richardson Carbon	Electronic Characterization of Carbon Black
Spring 1996	Best Symposium	L. Evans and S. Mowdood	J. M. Huber, Pirelli	Tire Traction
Fall 1996	Best Paper –Original Contribution	M. D. Ellul	Advanced Elastomers	Novel Thermoplastic Elastomer Vulcanizated Exhibiting Superior Low Temperature Performance and Elasticity

MEETING	AWARD	AUTHORS	AFFILIATION	PAPER TITLE
Fall 1996	Best Paper —Review	M. E. Wood and M. J. Recchio	Zeon Chemical	A Comparison of Improved HNBR Compounds versus Other Automotive Sealing Materials
Fall 1996	Honorable Mention —Original Contribution	A. A. Galuska, R. R. Poulter, and K. O. McElrath	Exxon	AFM Force Modulation Microscopy of Immiscible Elastomer Blends: Morphology, Fillers and Cross-Linking
Fall 1996	Honorable Mention —Review	M. C. Phair and T. Wampler	CDS	Analysis of Rubber Materials by Pyrolysis-Gas Chromatography
Fall 1996	Best Symposium	B. Rogers	Goodyear	Truck Tire Treadwear and Retreading
Spring 1997	Best Paper —Original Contribution	W. J. Patterson, M.-J. Wang, T. A. Brown, and H. Moneypenny	Cabot	Carbon/Silica Dual Phase Filler to Tire Tread Compounds
Spring 1997	Best Paper —Review	A. A. McNeish and J. T. Byers	Degussa	Low Rolling Resistance Tread Compounds —Some Compounding Solutions
Spring 1997	Honorable Mention —Original Contribution	U. Goerl and J. Muenzenberg	Degussa	Investigation into the Chemistry of the TESPT Sulfur Chain
Spring 1997	Best Educational Symposium Paper	A. Niziolek	Bayer	Solution SBR–Technology and Applications
Fall 1997	Best Paper —Original Contribution	J. Boerio, Y. M. Tsai, and D. K. Kim	University of Cincinnati, Goodyear	Adhesion of Natural Rubber to Steel Substrates: The Use of Plasma Polymerized Primers
Fall 1997	Best Paper —Review	L. Pomini	Techint Compagnia	HDM Tangential Rotors Technical Features and Technological Aspects
Fall 1997	Honorable Mention —Original Contribution	X. Zhang and R. Whitehouse	Cabot	Compatibility of Carbon Blacks with Typical EPDM Extrusion Compounds: Phenomenon and Root Causes
Fall 1997	Honorable Mention —Review	N. Nakajima	University of Akron	Science of Mixing of Rubber
Fall 1997	Best Symposium	B. L. (Les) Lee and W. H. Waddell	Pennsylvania State University, Exxon	Deformation and Fracture of Rubber Composites
Spring 1998	Best Paper —Original Contribution	N. R. Dharmarajan, P. S. Ravishankar, W. C. Horn, and W. A. Lambert	Exxon	New EPDM Elastomer for Wire and Cable Applications
Spring 1998	Best Paper —Review	J. B. Class	Hercules	A Review of the Fundamentals of Crosslinking with Peroxides

MEETING	AWARD	AUTHORS	AFFILIATION	PAPER TITLE
Spring 1998	Honorable Mention —Original Contribution	R. F. Ohm, R. G. Vara, and T. M. Buckley	R. T. Vanderbilt, DuPont	Sulfur Cure System Development for EPDM Produced via Constrained Geometry Catalyst Technology
Spring 1998	Best Educational Symposium Paper	J. T. Byers	Degussa	Silane Coupling Agents for Enhanced Silica Performance
Spring 1998	Best Symposium	V. Kothari	Therm-O-Link	Wire and Cable Applications
Fall 1998	Best Paper —Original Contribution	D. R. Keller, L. Bryant, and J. Dewar	Bayer	Enhanced Viscosity EVM Elastomers for General Purpose Molded and Extruded Applications
Fall 1998	Best Paper —Review	D. R. Hansen and S. J. St. Clair	Shell	Styrenic Thermoplastic Elastomers
Fall 1998	Honorable Mention —Original Contribution	A. C. Balazs, C. Singh, and E. Zhulina	University of Pittsburgh	Modeling the Interactions Between Polymers and Clay Surfaces Through Self-Consistent Field Theory
Fall 1998	Best Symposium	A. J. Dias and D. G. Peiffe	Exxon	Polymer Blends
Spring 1999	Best Paper —Original Contribution	K. F. Castner	Goodyear	Improved Processing CIS-1, 4-Polybutadiene
Spring 1999	Best Paper —Review	A. W. Niziolek, R. H. Jones, and J. G. Neilsen	Bayer	Influence of Compounding Materials on Tire Durability
Spring 1999	Honorable Mention —Original Contribution	M. C. Bulawa and P. S. Ravishankar	Exxon	Reduction of Peroxide Usage In EPDM Coolant Hose Via Polymer Redesign
Spring 1999	Best Educational Symposium Paper	A. G. Ferradino	R. T. Vanderbilt	Zinc 2-Mercaptotoluimidazole (ZMTI)– Review of a Unique and Versatile Antioxidant Synergist
Spring 1999	Best Symposium	A. L. Tisler, D. A. Kotz, H. W. Young, and W. F. Cole	Exxon, DuPont, DSM Copolymer, Flexsys America	Educational Symposium on Specialty Elastomers and Protective Chemicals for Their Use
Fall 1999	Best Paper —Original Contribution	M. L. Kerns, Z. G. Xu, and S. Christian	Goodyear	Synthesis of Random, Low Vinyl SSBR Using Distributed Monomer Feed Systems
Fall 1999	Honorable Mention —Original Contribution	T. M. Buckley, R. G. Vara, and R. F. Ohm	DuPont Dow Elastomers, R. T. Vanderbilt	Antidegradant System Development for EPDM Produced via Constrained Geometry Catalyst Technology
Fall 1999	Best Paper —Review	J. R. Serumgard, M. Blumenthal	Rubber Manufacturers Association	Overview of Scrap Tire Management and Markets in the United States

MEETING	AWARD	AUTHORS	AFFILIATION	PAPER TITLE
Fall 1999	Best Symposium	M. Engelhardt	Hankook Tire America	Surface/Interface Analysis of Tires and Rubber Compounds
Spring 2000	Best Paper —Original Contribution	J. W. M. Noordermeer and H. J. H. Beelen	DSM Elastomers	Understanding the Influence of Polymer and Compounding Variations on the Extrusion Behaviour of EPDM Compounds II: The Effect of Controlled Long Chain Branching
Spring 2000	Honorable Mention —Original Contribution	W. H. Waddell, R. C. Napier, and R. R. Poulter	ExxonMobil Chemical	Improved Tread Compound Wet/Winter Traction Using Brominated Isobutylene-co-para-Methylstyrene
Spring 2000	Best Paper —Review	J. C. Moreland and D. E. Hall	Michelin Americas	Fundamentals of Rolling Resistance
Spring 2000	Best Educational Symposium Paper	E. Bakuniec	Continental General Tire	Tire Construction for Normal Non-Speed-Rated and High-Speed Rated Passenger Car Tires (OE, Replacement, All Season)
Spring 2000	Best Symposium	W. Hopkins	Bayer Rubber	Educational Symposium on Basic Tire Technology
Fall 2000	Best Paper —Original Contribution	W. V. Mars	Cooper	Cracking Energy Density as a Predictor of Fatigue Life under Multiaxial Conditions
Fall 2000	Honorable Mention —Original Contribution	A. N. Gent and M. Rassagi-Kashani	Goodyear	Energy Release Rate for a Crack in a Tilted Block
Fall 2000	Best Paper —Review	O. H. Yeoh, G. Pinter, and H. T. Banks	Lord, North Carolina State University	Compression of Bonded Rubber Blocks
Fall 2000	Honorable Mention —Review	S. Abdou-Sabet	Consultant	Fifty Years of Thermoplastic Elastomer Innovations
Fall 2000	Best Symposium	J. T. Bauman, O. H. Yeoh	Elastomer Engineering and Testing, Lord	Engineering Design of Rubber Components
Spring 2001	Best Paper —Original Contribution	J. Putman, M. Arai	Pratt and Whitney, Shin-Etsu Chemical	Perfluoroether Elastomer Seal Development
Spring 2001	Honorable Mention —Original Contribution	G. J. Pehlert, N. R. Dharmarajan, and P. S. Ravishankar	ExxonMobil Chemical	Blends of EPDM And Metallocene Plastomers for Wire And Cable Applications
Spring 2001	Best Paper —Review	M. Coughlin, R. Schnell, and S. Wang	DuPont Dow Elastomers	Perfluoroelastomers in Severe Environments: Properties, Chemistry and Applications
Spring 2001	Best Symposium	V. Kothari	Therm-O-Link	Wire And Cable Applications
Fall 2001	Best Paper —Original Contribution	W. S. Fulton	Rhodia Industrial Specialties	Tyre Cord Adhesion–Interface Morphology and the Influence of Cobalt

APPENDIX 4

MEETING	AWARD	AUTHORS	AFFILIATION	PAPER TITLE
Fall 2001	Honorable Mention– Original Contribution	J. B. Putman and M. C. Putman	Tech Pro	An Improved Method for Measuring Filler Dispersion of Uncured Rubber
Fall 2001	Best Paper –Review	A. J. M. Sumner, R. Engelhausen, and J. Trimbach	Bayer	Polymer Developments to Improve Tire Life and Fuel Economy
Fall 2001	Best Symposium	G. T. Burns, J. E. Mark, T. A. Okel, and S. Sumimura	Dow Corning, University of Cincinnati, PPG	Silicone Rubber
Spring 2002	Best Paper –Original Contribution	G. M. Brown and A. D. Westwood	ExxonMobil Chemical	Application of Field Emission Scanning Electron Microscopy to Microanalysis of Elastomers
Spring 2002	Best Paper –Review	N. Yerina and S. Magonov	Digital Instruments/ Veeco Metrology	Atomic Force Microscopy in Analysis of Rubber Materials
Spring 2002	Best Educational Symposium Paper	H. D. Luginsland	Degussa AG	Chemistry and Physics of Network Formation in Silica-Silane-Filled Rubber Compounds: A Review on the Chemistry and the Reinforcement of the Silica-Silane Filler System
Spring 2002	Best Symposium	A. H. Tsou, W. H. Waddell, and T. W. Zerda	ExxonMobil Chemical	Microscopic and Spectroscopic Imaging of Rubber Compounds
Fall 2002	Best Paper –Original Contribution	I. Duvdevani, A. H. Tsou, and S. Datta	ExxonMobil Chemical	BIIR Green Strength Enhancement by Semicrystalline Specialty Propylene Elastomers
Fall 2002	Best Paper –Review	R. D. Stevens	DuPont Dow Elastomers	Long Term Heat Aging of Various Fluoroelastomers
Fall 2002	Best Symposium	B. DeMarco	Zeon Chemical	Heat Aging of Elastomers Under-the-Hood
Spring 2003	Best Paper –Original Contribution	C. Wrana, C. Fischer, and V. Haertel	Bayer Continental	Analysis of Network Structures in Filled Elastomers by Amplitude Dependent Measurements under Mono- and Bimodal Sinusoidal Deformation
Spring 2003	Honorable Mention –Original Contribution	J. Puskas and Y. Chen	University of Western Ontario	Novel Thermoplastic Elastomers for Biomedical Applications
Spring 2003	Best Paper –Review	D. Schwarz and D. Askea	Smithers Scientific Services	A Fundamental Review of Cut & Chip Testing for OTR Tread Compounds
Spring 2003	Honorable Mention –Review	F. Ignatz-Hoover and B. To	Flexsys America	Softening NR Compounds: A Comparison of Methods Used by the Industry to Increase Productivity of NR Compounds by Various Methods of Softening
Spring 2003	Best Symposium	M. Engelhardt	Yokohama	Medical Applications & Environmental Issues

MEETING	AWARD	AUTHORS	AFFILIATION	PAPER TITLE
Fall 2003	Best Paper –Original Contribution	N. Dharmarajan, M. G. Williams, and S. Datta	ExxonMobil Chemical	Specialty Elastomer Modifiers with Isotactic Propylene Crystallinity for Polypropylene Modification
Fall 2003	Honorable Mention –Original Contribution	A. K. Bhowmick and S. Sadhu	Indian Institute of Technology	Acrylonitrile Butadiene Rubber Based Nanocomposites: Preparation and Mechanical Properties
Fall 2003	Best Paper –Review	J. A. Kuczkowski	Consultant	Things I Learned about PPDs at My Mentor's Knee and Other Places
Fall 2003	Honorable Mention– Review	A. Sarkar, J. C. Dellamonte, and J. M. Caruthers	Purdue University	An Analysis of the Microscopic Deformation Field in Rubbers Filled with Nano-Particles
Fall 2003	Best Symposium	V. Kothari	Therm-O-Link	Aging and Stabilization of Rubber and Plastics
Spring 2004	Best Paper	J. Baldwin	Ford	Effects of Nitrogen Inflation on Tire Aging and Performance
Spring 2004	Honorable Mention	J. T. Bauman	Elastomer Engineer-ing and Testing	Calculation Methods for Spherical Rubber Bearings, Part II. Axial Forces
Spring 2004	Best Educational Symposium Paper	G. Burrowes	Goodyear	Applications–Power Transmission Products
Spring 2004	Best Symposium	M. R. Rodgers and W. H. Waddell	ExxonMobil Chemical	Educational Symposium: Elastomers and Their Applications
Fall 2004	Best Paper	R. J. Pazur, L. Ferrari, and E. Campomizzi	Lanxess	Compounding for Improved Physical and Heat Aging Properties in HXNBR
Fall 2004	Honorable Mention	W. S. Fulton	OMG UK, Ltd.	Tyre Cord Adhesion–How the Source of Zinc Can Influence the Structure of the Bonding Interface
Fall 2004	Best Symposium	V. Kothari	Therm-O-Link	Heat-Resistant Elastomers
Spring 2005	Best Paper	G. L. Wilkes, D. Klinedinst, J. P. Sheth, E. Yilgor, I. Yilgor, and F. L. Beyer	Virginia Polytechnic, Koc University, and U.S. Army Research Laboratory	Structure-Property Behavior of New Segmented Polyurethanes and Polyureas without Use of Chain Extenders
Spring 2005	Best Educational Symposium Paper	R. Costin and S. K. Henning	Sartomer	Fundamentals of Curing Elastomers with Peroxides and Coagents
Spring 2005	Best Symposium	S. K. Mowdood and W. Waddell	Consultant, Exxon-Mobil Chemical	Tire Performance, Durability and the Tread Act
Fall 2005	Best Paper	M. R. Gurvich	United Technologies	On Challenges in Structural Analysis of Rubber Components: Frictional Contact
Fall 2005	Best Symposium	A. L Tisler	Cytec Industries	Polyurethanes

MEETING	AWARD	AUTHORS	AFFILIATION	PAPER TITLE
Fall 2006	Best Paper	T. Hogan, Y. Yan, W. L. Hergenrother, and D. Lawson	Bridgestone Americas	Study of Lithiated Thioacetals as Initiators for Living Anionic Polymerization of Butadiene or Butadiene and Styrene. Polymerization and Compounded Polymer Properties
Fall 2006	Best Symposium	V. Kothari	Therm-O-Link	Advances in Rubber Technology
Fall 2007	Best Paper	R. P. Quirk and M. Ocampo	University of Akron	Anionic Synthesis of Trialkoxysilyl-Functionalized Polymers
Fall 2007	Best Symposium	V. Kothari	Therm-O-Link	Aging, Degradation and Stabilization of Rubber and Plastics
Fall 2008	Best Paper	W. Liang, S. Tang, R. Vara, C. Bette, T. Clayfield, S. Watson, and S. Daniel	Dow Chemical Elastomers R&D	Molecular Structure and Processing Characteristics of a Granular Gas-phase Metallocene EPDM Rubber
Fall 2008	Honorable Mention	W. Waddell, C. Napier, and D. Tracey	ExxonMobil Chemical	Nitrogen Inflation Of Tires
Fall 2008	Best Symposium	W. Waddell, and S. Mowdood	Consultant, Exxon-Mobil Chemical	Advances in Tire Technology

APPENDIX 5

Established in 1948, the 25-Year Club Committee is an honorary organization of Rubber Division members who have worked in the rubber
industry for at least twenty-five years. The club oversees and provides
direction regarding division activities related to the semiannual 25-Year
Club Luncheon and Reception. Jim D'Ianni was the club's first official
chairman in 1964, and the club adopted by-laws in 2003. Ralph Graff,
who took the chairmanship in 1981, is the recognized "father" of the
organization. Mementos, significant of the living member with the most
years of industry service, were first awarded in 1952.

DATE	ATTENDANCE	LOCATION	MASTER OF CEREMONIES
Fall 1948	134	Chicago	H. A. Winkelmann
Spring 1949	124	Detroit	Wm. Nelson
Fall 1949	136	Boston	J. Burer
Spring 1950	—	Atlantic City	J. Coe
Fall 1950	147	Cleveland	R. P. Dinsmore
Spring 1951	108	Washington	H. L. Fisher
Fall 1951	175	New York	A. A. Somerville
Spring 1952	139	Cincinnati	A. C. Eide
Fall 1952	155	Buffalo	H. F. Van Valkenburg
Spring 1953	196	Boston	W. E. Kavanagh

DATE	ATTENDANCE	LOCATION	MASTER OF CEREMONIES
Fall 1953	140	Chicago	H. A. Winkelmann
Spring 1954	150	Louisville	A. Brandt
Fall 1954	222	New York	B. R. Silver
Spring 1955	162	Detroit	E. J. Kvet
Fall 1955	198	Philadelphia	A. H. Nelson
Spring 1956	194	Cleveland	F. W. Stavely
Fall 1956	170	Atlantic City	E. R. Bridgewater
Spring 1957	160	Montreal	N. S. Grace
Fall 1957	202	New York	W. O. Hamister
Spring 1958	123	Cincinnati	H. S. Karch
Fall 1958	137	Chicago	C. M. Baldwin
Spring 1959	141	Los Angeles	E. G. Partridge
Fall 1959	200	Washington	C. A. Bartle
Spring 1960	115	Buffalo	C. H. Peterson
Fall 1960	200	New York	J. M. Ball
Spring 1961	124	Louisville	H. O. Geroge
Fall 1961	142	Chicago	B. W. Lewis
Spring 1962	187	Boston	O. J. Brown Jr.
Fall 1962	193	Cleveland	A. E. Juve
Spring 1963	163	Toronto	N. S. Grace
Fall 1963	213	New York	J. Breckley
Spring 1964	183	Detroit	S. M. Caldwell
Fall 1964	166	Chicago	B. W. Lewis
Spring 1965	163	Miami Beach	R. L. Holmes
Fall 1965	176	Philadelphia	G. Wyrough
Spring 1966	150	San Francisco	W. D. Good
Fall 1966	230	New York	E. Kern
Spring 1967	174	Montreal	D. Huggenberger
Fall 1967	169	Chicago	H. Shetler
Spring 1968	254	Cleveland	D. Doherty
Fall 1968	180	Atlantic City	S. Martin
Spring 1969	187	Los Angeles	C. Kuhn
Fall 1969	—	Buffalo	C. H. Peterson

DATE	ATTENDANCE	LOCATION	MASTER OF CEREMONIES
Spring 1970	206	Washington	Dr. L. A. Wood
Fall 1970	147	Chicago	Stan Choate
Spring 1971	151	Miami Beach	D. Reneau
Fall 1971	220	Cleveland	R. F. Wolf
Spring 1972	206	Boston	J. Hussey
Fall 1972	162	Cincinnati	F. Gage
Spring 1973	157	Detroit	Dave A. Sherman
Fall 1973	185	Denver	Carl P. Mullen
Spring 1974	207	Toronto	Dr. N. Grace
Fall 1974	185	Philadelphia	George Wyrough
Spring 1975	217	Cleveland	Tom Pollard
Fall 1975	157	New Orleans	Robert Sontag
Spring 1976	144	Minneapolis	Fred Fisher
Fall 1976	193	San Francisco	Ralph Hickox
Spring 1977	188	Chicago	Elroy Ekebus
Fall 1977	229	Cleveland	Dr. E. Gruber
Spring 1978	192	Montreal	Jack Carr
Fall 1978	226	Boston	Robert Loveland
Spring 1979	206	Atlanta	Thomas. W. Elkin
Fall 1979	219	Cleveland	Robert H. Gerster
Spring 1980	169	Las Vegas	A. J. Hawkins Jr.
Fall 1980	152	Detroit	William J. Simpson
Spring 1981	161	Minneapolis	Clyde I. Hause
Fall 1981	239	Cleveland	William Ferguson
Spring 1982	196	Philadelphia	Thomas N. Loser
Fall 1982	162	Chicago	Robert R. Kann
Spring 1983	171	Toronto	Jack Carr
Fall 1983	179	Houston	Wes Muller
Spring 1984	173	Indianapolis	Lewis E. Roccaforte
Fall 1984	164	Denver	Frank Burton
Spring 1985	165	Los Angeles	Lawrence Peterson
Fall 1985	201	Cleveland	Robert H. Gerster
Spring 1986	135	New York	Jack Peto

DATE	ATTENDANCE	LOCATION	MASTER OF CEREMONIES
Fall 1986	133	Atlanta	Alfred Cobbe
Spring 1987	128	Montreal	Pierre Venne
Fall 1987	157	Cleveland	Darrel Cox
Spring 1988	100	Dallas	Jack Stewart
Fall 1988	110	Cincinnati	Marvin Coulter
Spring 1989	100	Mexico City	Oscar Moreno Valdes
Fall 1989	150	Detroit	Walden D. Wilson
Spring 1990	110	Las Vegas	John T. Moriarty
Fall 1990	126	Washington	Ralph S. Graff
Spring 1991	80	Toronto	Tom A. Crooks
Fall 1991	83	Detroit	Harold E. Trexler
Spring 1992	65	Louisville	Robert Miller
Fall 1992	104	Nashville	Fran Watkins
Spring 1993	83	Denver	Henry Tramutt
Fall 1993	104	Orlando	Roger T. Read
Spring 1994	77	Chicago	Bert C. Vandermar
Fall 1994	93	Pittsburgh	Ralph S. Graff
Spring 1995	88	Philadelphia	Ralph S. Graff
Fall 1995	108	Cleveland	Francis M. O'Connor
Spring 1996	84	Montreal	Jacques Aumais
Fall 1996	93	Louisville	Thomas J. Roe
Spring 1997	92	Anaheim	Ed Breslin
Fall 1997	128	Cleveland	Walter C. Warner
Spring 1998	105	Indianapolis	Les M. Kerr
Fall 1998	110	Nashville	Les C. Jacobson Jr.
Spring 1999	74	Chicago	Wesley H. Whittington
Fall 1999	92	Orlando	John R. Deputy
Spring 2000	69	Dallas	Deborah L. Banta
Fall 2000	72	Cincinnati	Richard J. Hupp
Spring 2001	65	Providence	Jim Graham
Fall 2001	68	Cleveland	Charles Rader
Spring 2002	60	Savannah	John Howard
Fall 2002	64	Pittsburgh	Gerald Williams

DATE	ATTENDANCE	LOCATION	MASTER OF CEREMONIES
Spring 2003	45	San Francisco	John Moriarty
Fall 2003	70	Cleveland	John Long
Spring 2004	48	Grand Rapids	Robert Pett
Fall 2004	57	Columbus	Charles Rader
Spring 2005	40	San Antonio	Donovan Parsons
Fall 2005	70	Pittsburgh	Ed Lohr
Spring 2006	58	Akron	Ben Kastein
Fall 2006	53	Cincinnati	John Deputy
Spring 2007	53	Akron	Charles Rader
Fall 2007	66	Cleveland	Isaac Shilad
Spring 2008	37	Dearborn	Robert Pett
Fall 2008	55	Louisville	Bruce Milligan

25-Year Club Memento Winners

DATE	NAME	AFFILIATION AT TIME OF AWARD	YEARS IN THE RUBBER INDUSTRY
Spring 1952	B. W. Henderson	American Cyanamid	—
Fall 1952	W. E. Kavanagh	Goodyear Tire & Rubber	—
Spring 1953	Harold Fuller	Pequanoc Soft Rubber	—
Fall 1953	F. Malin	Bell Laboratories	—
Spring 1954	P. E. Cholet	Allied Chemicals and Dye	—
Fall 1954	Frank Baker	Landers Corporation	—
Spring 1956	R. R. Olin	Olin Laboratories	—
Fall 1956	H. Underwood	Minnesota Mining and Mfg.	—
Spring 1956	A. W. Holmberg	Naugatuck Chemical	—
Fall 1956	Clinton Schultz	Schenuit Rubber	—
Spring 1957	C. B. Copeman	*Rubber Journal* (England)	—
Fall 1957	H. Muehlstein	H. Muehlstein and Company	—
Spring 1958	George Kratz	General Latex and Chemical	—
Fall 1958	J. E. Warrill	Carlisle Tire and Rubber	—
Spring 1959	Paul Vancleef	Johns-Manville	—
Fall 1959	Irving Laurie	Laurie Rubber Reclaiming	—

DATE	NAME	AFFILIATION AT TIME OF AWARD	YEARS IN THE RUBBER INDUSTRY
Spring 1960	Gordon Holmes	St. Lawrence Chemical	—
Fall 1960	C. E. Bauer	Hercules Tire and Rubber	—
Spring 1961	C. R. Johnson	Spencer Products	—
Fall 1961	G. S. Whitby	University of Akron	—
Spring 1962	J. M. Bierer	Boston Woven Hose and Rubber	—
Fall 1962	W. D. Good	American Rubber Mfg.	—
Spring 1963	Harry Atwater	Consultant	—
Fall 1963	J. C. Wood	R. T. Vanderbilt Company	—
Spring 1964	R. P. Dinsmore	Goodyear Tire & Rubber	—
Fall 1964	S. S. Skelton	S. S. Skelton	—
Spring 1965	A. Kemple	Rex-Hide	—
Fall 1965	F. Dannerth	Consultant	—
Spring 1966	E. A. Murphy	Dunlop, Ltd. (England)	—
Fall 1966	W. S. Gibbons	Uniroyal (retired)	—
Spring 1967	S. Mclane	British Rubber	—
Fall 1967	L. Healey	Consultant	—
Spring 1968	W. Whittaker	Retired	—
Fall 1968	R. Spreat	Acme Hazilton Mfg. (retired)	—
Spring 1969	S. Collier	UCLA	—
Fall 1969	Dr. Arnold Smith	B. F. Goodrich (retired)	56
Spring 1970	Dr. W. R. Hucks	Consultant	54
Fall 1970	Edward Meyer	Herron and Meyer	58
Spring 1971	Alfred Stein	Muehlstein (retired)	56
Fall 1971	Alvin C. Peterjohn	Uniroyal (retired)	56
Spring 1972	Morris Omansky, A. C. Eide	American Zinc (retired)	56
Fall 1972	S. J. Pike	Pike and Alcan	56
Spring 1973	Harvey Doering, Warren C. Carter	R. T. Vanderbilt (retired), Kenrich	54
Fall 1973	George Wyrough	Wyrough and Loser (retired)	53
Spring 1974	Art Briant	Retired	64

DATE	NAME	AFFILIATION AT TIME OF AWARD	YEARS IN THE RUBBER INDUSTRY
Fall 1974	Clyde Hoover	Hoover Hanes (retired)	54
Fall 1974	Art Ross	Am. Biltrite (retired)	54
Spring 1975	Henry Rose	Muehlstein (retired)	58
Fall 1975	Bernie Capen	(Retired)	57
Spring 1976	John Hajicek	Twin City Rubber	58
Spring 1976	Ralph Robinson	Robinson Rubber	58
Fall 1976	Charles Baldwin	Baldwin	56
Spring 1977	Edward Goodman	Retired	58
Fall 1977	Dr. Harlan Trumbull	B. F. Goodrich (retired)	58
Spring 1978	O. M. (Bill) Hayden	Dupont De Nemours	59
Fall 1978	Donald Wright	B. F. Goodrich (retired)	59
Spring 1978	W. Walker	B. F. Goodrich	60
Fall 1979	F. Yoder	B. F. Goodrich	67
Spring 1980	Leo Dete	Carlisle Rubber (retired)	58
Fall 1980	Richard K. Patrick	Vulplex (retired)	58
Spring 1981	C. Walz Durkee	Atwood (retired)	64
Fall 1981	I. Patterson	Goodyear (retired)	61
Spring 1982	Gavin A. Taylor	Exxon Chemical (retired)	60
Fall 1982	Joseph J. Tumpeer	Wishnick-Tumpeer	62
Spring 1983	George Eykamp	Midstates Rubber Prod.	54
Fall 1983	Cecil Draper	Geneva Rubber (retired)	55
Spring 1984	William H. Lussie	R. T. Vanderbilt (retired)	52
Fall 1984	Philip M. Earhart	Gates Rubber (retired)	57
Spring 1985	Howard Wiley	Gates Rubber (retired)	59
Fall 1985	John M. Walsh	Firestone Xylos Div. (retired)	57
Spring 1986	Bryant Ross	Pennwalt (retired)	—
Fall 1986	William Haney	Kirkmill Rubber	57
Spring 1987	Robert H. Gerster	Goodyear (retired)	—
Fall 1987	Robert Juve	Mohawk Rubber (retired)	54
Spring 1988	Ross C. Whitmore	Better Monkey Grip (retired)	65
Fall 1988	Ralph Wheaton	St. Joe Zinc	52

DATE	NAME	AFFILIATION AT TIME OF AWARD	YEARS IN THE RUBBER INDUSTRY
Spring 1989	A. H. Woodward	Dupont (retired)	54
Fall 1989	Edward Aurt	U.S. Rubber (retired)	60
Spring 1990	Henry Kehe	B. F. Goodrich (retired)	55
Fall 1990	Robert Babitt	R. T. Vanderbilt (retired)	58
Spring 1991	Herman Schroeder, Art Moline, Richard Silver	Dupont (retired), Consultant (retired), Akrochem	53
Fall 1991	Stephen T. Semegen	Valley Rubber Mixing	51
Spring 1992	R. Van Patten-Steiger	Van Grace Associates (retired)	56
Fall 1992	Pasquale A. Uva	Avon Sale (retired)	63
Spring 1993	Walton Wilson	Gates Rubber (retired)	56
Fall 1993	Alvert Benckel	Beloit-Manhattan (retired)	57
Spring 1994	Irwin O. Nejdt	Ideal Roller (retired)	57
Fall 1994	Ernie L. Puskas Sr.	E. L. Puskas	54
Spring 1995	James K. Langkamer	Acme Hamilton Mfg.	53
Fall 1995	F. Merle Galloway	H. K. Porter	62
Spring 1996	Walt Warner	*Rubber World*	53
Fall 1996	Angelo A. Lucia	E. L. Puskas	54
Spring 1997	Frank Steitz	American Chemet	61
Fall 1997	James D'Ianni	Goodyear (retired)	62
Spring 1998	Cesareo A. Gonzalez	Gonzalez Cano Y. Cia	54
Fall 1998	Don Bordenkirch	Crossfield Rubber (retired)	57
Spring 1999	Bert E. Vandermar, Ralph S. Graff	Van-Fron, Dupont (retired)	53
Fall 1999	M. E. "Butch" Noah	American Chemet	57
Spring 2000	Henry Richmond	U.S. Rubber (retired)	58
Fall 2000	Edward L. Nazar	Nazar Rubber	54
Spring 2001	Joseph D. Guisto	Acushnet Rubber	52
Fall 2001	W. G. Desai	Indian Rubber Manufacturers Research Association	58

DATE	NAME	AFFILIATION AT TIME OF AWARD	YEARS IN THE RUBBER INDUSTRY
Spring 2002	Ralph Hoy	Exxon (retired)	56
Fall 2002	Lois Brock	GenCorp (retired)	56
Spring 2003	William Scheuermann	Paroo (retired)	62
Fall 2003	Ray Parker	Goodall (retired)	63
Spring 2004	James Graham	Dupont (Retired)	53
Fall 2004	Ted J. Van Veggel	Airboss Rubber Compounding	54
Spring 2005	Manu Patel	Apar Industries, Ltd.	53
Fall 2005	Jim Proudfit	J&R Consulting Services, LLC	53
Spring 2006	Ben Kastein	Firestone Tire (retired)	66
Fall 2006	Al Blum	Chemintertech Associates	59
Spring 2007	Bernard Hugus	Polyonics Rubber	59
Fall 2007	James McCarthy	Chemrep	53
Spring 2008	Frank Kelley	University of Akron (retired)	56
Fall 2008	Richard Eykamp	Mid-States Rubber Products	52

NOTES

CHAPTER 1. KING OF THE JUNGLE

1. A publication of the Académie Royale des Sciences.

2. "Definition(s) of Rubber," *Rubber & Plastics News*, August 13, 1984, 19.

3. "The Search for a Rubber Solvent," *Rubber & Plastics News*, August 13, 1984, 24.

4. Austin Coates, *The Commerce in Rubber: The First 250 Years* (Oxford: Oxford University Press, 1987), 29.

5. The name Para also became a common prefix used with rubber. It referred to rubber from Brazil's Pará province, which was the province located at the mouth of the Amazon River from where a great majority of rubber was exported to points all over the world.

6. "Search for a Rubber Solvent," 24.

7. Paul E. Hurley, "History of Natural Rubber," in *Rubber World Special 75th Anniversary Issue of the Rubber Division*, ed. Don Smith (Akron, Ohio: Lippincot, 1984), 10.

8. Ibid.

CHAPTER 2. GENESIS

1. Marston T. Bogert, "American Chemical Societies," *Journal of the American Chemical Society*, February 1908, 30(2): 163–82.

2. Today, after a name change and a merger, it is part of the Drexel College of Medicine in Philadelphia.

3. Bogert, "American Chemical Societies."

4. Bodley was born on December 7, 1831, in Cincinnati, Ohio, but eventually migrated to Macon, Georgia, and Wesleyan Female College, where she earned her bachelor of arts degree in 1849. Wesleyan Female College was the

first institution in the world chartered to grant degrees to women. She likely never thought of herself as a trailblazer, but whether or not she realized it, Bodley would become a personal missionary of that school's mantra for the rest of her life.

In 1865, she moved to Philadelphia and became not only the first female professor of chemistry at the Women's Medical College of Pennsylvania but also, eventually, its first female dean. However, that wasn't her last stop on the feminist trail. The reply letter to the *American Chemist* provided the catalyst for the meeting of the chemists in Northumberland, which enabled the seed to be planted, ultimately, for the establishment of the American Chemical Society in 1876. As a result, and largely because of her work on the centennial celebration in honor of Joseph Priestley's discovery of oxygen, in 1876 she was elected a charter member of the American Chemical Society. She later resigned her membership in the ACS in protest after an 1880 dinner in Boston where women were barred from participation and "Chemistry festivities" afterward included a performance of antifemale songs and poems.

5. R. S. Bates, *Scientific Societies in the United States*, 3rd ed. (Boston: Massachusetts Institute of Technology Press, 1965), 99–100.

6. Ibid.

7. Bogert, "American Chemical Societies," 169.

8. George W. Knepper, *New Lamps for Old: One Hundred Years of Urban Higher Education at the University of Akron* (Akron, Ohio: University of Akron Press, 1970), 37.

CHAPTER 3. LEAP OF FAITH

1. William Woodruff, "Growth of the Rubber Industry of Great Britain and the United States," *Journal of Economic History* 15(4): 376–391.

2. U.S. Department of Transportation, Federal Highway Administration, Highway Statistics Summary, 2003, http://www.fhwa.dot.gov/pubstats.html/.

3. P. W. Barker, *Rubber Statistics, Trade Promotion Series, No. 181*, United States Department of Commerce (Washington, D.C.: Government Printing Office, 1938). Rubber Statistics, 1900–1937: Production, Absorption, Stocks, and Prices.

4. H. T. Collings, "The Relation of the Automobile Industry to International Problems of Oil and Rubber," *Annals of the American Academy of Political and Social Science* 116 (1924): 254–258.

5. L. E. Carlsmith, "The Economic Characteristics of Rubber Tire Production" (Ph.D. diss., Columbia University, 1935).

6. Hurley, "History of Natural Rubber," 9.

7. Daniel Nelson, "Mass Production and the U.S. Tire Industry," *Journal of Economic History* 47 (June 1987): 329–39.

8. R. R. Resor, "Rubber in Brazil: Dominance and Collapse, 1876–1945," *Business History Review* 51 (Autumn 1977): 341–66.

9. Barker, *Rubber Statistics, Trade Promotion Series.*

10. G. Wright, "The Origins of American Industrial Success, 1879–1940," *American Economic Review* (1990), 80(4): 653.

11. M. G. Blackford and K. Austin Kerr, *B. F. Goodrich: Tradition and Transformation, 1870–1995* (Columbus: Ohio State University Press, 1996), 29.

12. Don Smith, "RW's Evolution Mirrored Industry's Needs," *Rubber World*, October 1989, 21.

13. Pontianak is the name of a grade of natural rubber from Indonesia, named after the city in Borneo, Indonesia.

14. Benjamin Kastein Jr., "Historical Highlights of the Rubber Division ACS," *Rubber Division 75th Anniversary, 1909–1984, Rubber World*, 1984, 43.

15. G. W. Knepper, *Summit's Glory: Sketches of Buchtel College and the University of Akron* (Akron, Ohio: University of Akron Press, 1990), 39.

16. Arnold H. Smith, tape-recorded interview by Herbert A. Endres, April 7, 1966, Science and Technology Library, University of Akron, Akron, Ohio.

CHAPTER 4. IN THE SHADOW

1. Woodruff, "Growth of the Rubber Industry," 376–91.

2. Samuel E. Horne Jr., "History of Synthetic Rubber," *Rubber Division 75th Anniversary, Rubber World*, 1984, 3.

3. Waldo N. Semon, *Chemical Engineering News*, 1943, 1613.

4. Horne, "History of Synthetic Rubber," 3.

5. O. Scott, B. Mulligan, J. Del Gatto, and K. Allison, eds., "The Division of Rubber Chemistry: Catalyst of an Industry," *Rubber World*, October 1966, 155(1): 70.

6. Nelson, "Mass Production and the U.S. Tire Industry," 331–32.

7. Ray P. Dinsmore, tape-recorded interview by Herbert A. Endres, February 8, 1965, Science and Technology Library, University of Akron, Akron, Ohio.

8. Loren B. Sebrell, tape-recorded interview by Herbert A. Endres, September 17, 1965, Science and Technology Library, University of Akron, Akron, Ohio.

9. Kastein, "Historical Highlights of the Rubber Division ACS," 42–46.

10. Smith interview by Endres.

11. Ralph F. Anderson, "Looking Back," lecture, Akron Rubber Group 50th Anniversary meeting, Akron, Ohio, February 15, 1978.

12. Coates, *Commerce in Rubber*, 239.

13. Resor, "Rubber in Brazil," 341–66.

14. Scott et al., eds., "Division of Rubber Chemistry," October 1966, 83.

15. The Bierer-Davis Oxygen Bomb isn't a bomb in the sense of a weapon of mass destruction. It is an industry-wide standard test to determine aging properties of rubber.

CHAPTER 5. TECHNICALLY SPEAKING

1. Before World War II, chewing gum (a.k.a. bubble gum) was made primarily of chicle, a latex sap that comes from the sapodilla tree, native to Central America. Today, chemists have learned how to make artificial gum bases to replace chicle. These gum bases are essentially synthetic rubbers that have the same temperature profile as chicle. Source: http://www.howstuffworks.com/.

2. The paper was also presented July 5–8, 1999, at the Seventh International Conference on Scintometrics and Informetrics, Colima, Mexico.

3. O. Scott, B. Mulligan, J. Del Gatto, and K. Allison, eds., "The Division of Rubber Chemistry: Catalyst of an Industry, Part II," *Rubber World*, November 1966, 90.

4. John M. Bierer, tape-recorded interview by Herbert A. Endres, April 20, 1966, Newton, Mass., Science and Technology Library, University of Akron, Akron, Ohio.

5. Tom H. Rogers, "Rubber Chemistry and Technology: Fifty Years of Achievement," *Rubber Chemistry and Technology*, 1978, 50(5): G106–G110.

CHAPTER 6. GROUP THERAPY

1. "Dow Jones Indexes," *Dow Jones Indexes*, February 29, 2008, Dow Jones Company, http://www.djindexes.com/mdsidx/index.cfm?event=showavgstats #n04/.

2. The society was also planning a large fiftieth anniversary celebration for September 1926 in Northumberland, Pennsylvania, the resting place of Joseph Priestley.

3. Dinsmore interview by Endres, February 8, 1965.

4. Ibid.

5. E. V. Osberg, "History of the Division of Rubber Chemistry of the American Chemical Society," *Rubber Chemistry and Technology*, 1940, 13: 1–10.

6. R. H. Crossley and Ralph F. Wolf, "A History of the Akron Rubber Group, Inc., 1928–1948," *Akron Rubber Group, Inc.: 50th Anniversary, 1928–1978*. 1978.

7. Official division minutes indicate that historian Herbert A. Endres addressed the new Los Angeles Group on their birth on January 20, 1928, twenty-six days before the Akron Rubber Group. For issues of practicality, they were all formed in early 1928, simultaneously.

8. The Boston Group held a preliminary meeting on May 9, 1928, and this informational meeting with 285 attendees and "many well-known rubber men from New England" precipitated the official organizational meeting.

9. "Associations, Societies Push Rubber Industry Progress," *Rubber World*, October 1989, 127.

10. Fred Pedersen, telephone interview by author, March 2008.

11. Ed Miller, interview by author, September 2007, Akron, Ohio.

12. George K. Hinshaw, tape-recorded interview by Herbert A. Endres, August 19, 1967, in New Concord, Ohio, Science and Technology Library, University of Akron, Akron, Ohio.

CHAPTER 7. DIVINE DISCOVERY

1. Arnold M. Collins, tape-recorded interview by DuPont Public Relations Department, May 1981, Science and Technology Library, University of Akron, Akron, Ohio.

2. Whitmore became president of the American Chemical Society in 1938.

3. Herman E. Schroeder, "Nieuwland and Neoprene: A Retrospective Reassessment," *Rubber World*, 1980, 18(6): 40.

4. Ernest R. Bridgewater, tape-recorded interview by Herbert A. Endres, October 15, 1965, Science and Technology Library, University of Akron, Akron, Ohio.

5. Arnold M. Collins, "The Discovery of Polychloroprene," lecture, Charles Goodyear Medal Banquet, Detroit, May 3, 1973.

6. Bridgewater interview by Endres.

7. Harlan L. Trumbull, tape-recorded interview by Herbert A. Endres, January 16, 1965. Science and Technology Library, University of Akron, Akron, Ohio.

8. James D. D'Ianni, "Highlights of Progress in Rubber and Chemicals," lecture, Winter Meeting of the Akron Rubber Group Fiftieth Anniversary, February 15, 1978.

9. Collins interview by DuPont.

CHAPTER 8. DÉJÀ VU

1. Hurley, "History of Natural Rubber," 10.

2. Frank A. Howard, *Buna Rubber: The Birth of an Industry* (New York: D. van Nostrand, 1947). Figures from U.S. Tariff Commission Report No. 6, September 1944.

3. *Rubber Statistics*, 1900–1937, by P. W. Barker (Washington, D.C.: U.S. Government Printing Office, 1938).

4. American Chemical Society, "National Historic Chemical Landmarks: United States Synthetic Rubber Program, 1939–1945," American Chemical Society, http://acswebcontent.acs.org/landmarks/landmarks/rbb/rbb_origin .html/.

5. Waldo Semon, tape-recorded interview by Division Historian Herbert A. Endres, January 19, 1966, Science and Technology Library, University of Akron, Akron, Ohio.

6. The company became Bayer in 1945.

7. Trumbull interview by Endres.

8. G. S. Whitby, C. C. Davis, and R. F. Dunbrook, *Synthetic Rubber: prepared under the auspices of the Division of Rubber Chemistry, American Chemical Society* (New York: John Wiley & Sons, 1954), 938.

9. Horne, "History of Synthetic Rubber," 5.

10. Howard, *Buna Rubber*, 72.

11. Royce J. Noble, "Rubber Technology Conference," *Rubber Age*, June 1938, 163.

12. Ibid.

CHAPTER 9. SUMMA CUM LAUDE

1. Hinshaw interview by Endres.

2. Bierer interview by Endres.

3. Rubber Division of American Chemical Society By-Law XV.

4. *Time*, March 29, 1941, http://www.time.com/time/magazine/article/ 0,9171,772964,00.html?iid=chix-sphere/.

5. Rubber Division of the American Chemical Society meeting minutes, September 11, 1942, University of Akron Archives.

6. Willis A. Gibbons, tape-recorded interview by Herbert A. Endres, April 14, 1966, Washington, D.C.

7. The reason there is no official record of Oenslager delivering a speech is because he didn't deliver one. However, division veterans say Oenslager wanted to speak about his family heritage, and he had discovered his ancestors were Native Americans. Since Charles Goodyear medalists are obligated to deliver a paper pertinent to the subject about which they received the medal in the first place, Oenslager was not allowed to speak—according to division folklore.

8. Ralph Graff, telephone interview by author, August 2007.

9. Tom H. Rogers, "Charles Goodyear Medalists and Charles Goodyear," *Rubber Chemistry and Technology*, May–June 1977, 50(2): G38–G41.

10. Ibid.

CHAPTER 10. CRITICAL COMMODITY

1. Glenn D. Babcock, *History of the United States Rubber Company* (Bloomington: Indiana University Press, 1966), 391.

2. William M. Tuttle Jr., "The Birth of an Industry: The Synthetic Rubber 'Mess' in World War II," in *Technology and Culture* (Baltimore: Johns Hopkins University Press, 1981), 22(1): 35–67.

3. Benjamin Kastein, interview by author, November 2007, Silver Lake, Ohio.

4. In June 1967, John McGavack, who had been instrumental in the maintenance of the Rubber Division Bibliography from 1940 to 1964, conducted an exhaustive study of the top one hundred contributors to the world's rubber literature between the years 1932 and 1966. His results were published in *Rubber Journal* and *Rubber Chemistry and Technology* and included technical articles published in the twenty-year time frame from authors in twenty-six nations. Likely to no one's surprise in the rubber industry, *RC&T* was the preferred forum for the more than twenty-two thousand references. A similar study was conducted in December 2000 by Wai Sin Tiew and K. Kaur and published by the *Malaysian Journal of Library and Information Science*. The Malaysian study yielded similar results, and *RC&T* topped the list as the most cited journal or serial publication in the field of rubber research. The study further stated, "In terms of frequency distribution of journals/serial citations, most of the serial publications (95.07%) received between one to ten citations. *Rubber Chemistry and Technology* was the most cited journal/serial title with 198 citations." It concluded, "Hence, this conforms to earlier studies, which indicate that rubber scientists depend highly on journals/serial literature."

5. Blackford and Kerr, *B. F. Goodrich*, 147–48.

6. John Collyer, tape-recorded interview by Benjamin Kastein Jr., February 12, 1979. Science and Technology Library, University of Akron, Akron, Ohio.

7. With the help of an arguable "Who's Who" team of U.S. synthetic rubber researchers, chemists, and producers, B. F. Goodrich and Hycar Chemical (Phillips) on June 5, 1940, at a press conference in New York's Waldorf-Astoria Hotel, announced a new tire made from a unique butadiene-acrylonitrile combination—the first tire made of synthetic rubber through American ingenuity and produced in the United States. The name of the new synthetic rubber was Ameripol, an appropriate combination of American and polymer.

8. The Rubber Development Corporation was formed primarily to maximize rubber production in Brazil.

9. John T. Wooley and Gerhard Peters, "The American Presidency Project," *Gerhard Peters*, March 30, 2008, 157, http://www.presidency.ucsb.edu/ws/index.php?pid=16236/.

10. *Encyclopedia Britannica*, 15th ed., s.v. "World War II."

11. Meyer previously rejected an offer of employment from Bernard Baruch to join Baruch's law firm.

12. Tuttle, "Birth of an Industry," 35–36.

13. Standard Oil of New Jersey was added to the prewar list of just four because of its access to patents on German Buna S synthetic rubber.

14. James D. D'Ianni, interview by author, June 11, 2007, Akron, Ohio.

15. Ernest T. Handley, tape-recorded interview by Benjamin Kastein Jr., 1980, Science and Technology Library, University of Akron, Akron, Ohio.

16. Tuttle, "Birth of an Industry," 39.

17. James D. D'Ianni, lecture, Washington Rubber Group and the Rubber Division, Washington, D.C., July 1979.

18. Maurice Morton, "A History of Synthetic Rubber," in *History of Polymer Science and Technology*, ed. Raymond B. Seymour (New York: Marcel Dekker, 1982), 2225–38.

19. Peter J. T. Morris, *The American Synthetic Rubber Research Program* (Philadelphia: University of Pennsylvania Press, 1989).

20. Ibid.

21. Dinsmore interview by Endres.

22. George S. Whitby, "A New Tetramethylbutadiene," *Rubber Chemistry and Technology*, 1928, 341.

23. Knepper, *New Lamps for Old*, 256.

24. James D. D'Ianni, lecture, Washington Rubber Group and the Rubber Division, Washington, D.C., July 1979.

25. Bradley Dewey, tape-recorded interview by Herbert A. Endres, 1965, Science and Technology Library, University of Akron, Akron, Ohio.

26. Melvin Lerner, "Pilot Plant and Laboratory Formally Dedicated at Akron," *Rubber Age*, July 1944, 11.

27. Minutes of the Rubber Division meeting, 1941–45, University of Akron Archives.

28. Morris, *American Synthetic Rubber Research Program*, 14.

CHAPTER 11. A RENAISSANCE

1. Louis D. Johnston and Samuel H. Williamson, "Nominal GDP: The Annual Real and Nominal GDP for the United States, 1789–Present," Economic History Services, 2004, http://www.eh.net/hmit/gdp/GDPsource.htm/.

2. Alan S. Milward, *War, Economy, and Society, 1939–1945* (Berkeley and Los Angeles: University of California Press, 1979).

3. *Funk and Wagnall's New Encyclopedia*, s.v. "Rosie the Riveter: Real Women Workers in World War II."

4. Barbara Boxer, "Women's History Timeline, 1900–1949," http://boxer.senate.gov/whm/time_3.cfm/.

5. U.S. Census Bureau, "Census of Population, 1950," http://factfinder.census.gov/servlet/.

6. Charles F. Phillips Jr., "Workable Competition in the Synthetic Rubber Industry," *Southern Economic Journal*, 1961, 28(2): 154–62.

7. Robert A. Solo, *Across the High Technology Threshold: The Case of Synthetic Rubber* (Norwood, Pa.: Norwood Editions, 1980).

8. In the fall of 1945, on President Truman's directive, a committee was established and chaired by William L. Batt to investigate long- and short-term requirements for natural and synthetic rubber and make recommendations.

The committee made many recommendations, many of which were adopted in the Rubber Act of 1948. But the Batt committee did not enact any legislation and was disbanded in 1947.

9. William A. Blanpied, "Inventing U.S. Science Policy," *Physics Today*, 1998, 51(2): 34.

10. Ibid., 36.

11. Lawrence R. Hafstad, "Science and Administration," *Public Administration Review*, 1951, 11(1): 10–16.

12. Conant and Compton had delivered keynote speeches in September 1939 at a Rubber Division meeting in Boston in recognition of the centennial of Charles Goodyear's discovery of vulcanization. Goodyear Tire & Rubber Company chairman P. W. Litchfield also delivered an address.

13. Paul R. Samuelson, "The U.S. Government Synthetic Rubber Program, 1941–1955: An Examination in Search of Lessons for Current Energy Technology Commercialization Projects," working paper, Energy Laboratory, Massachusetts Institute of Technology, Cambridge, 1976.

14. Norman A. Shepard, tape-recorded interview by Herbert A. Endres, October 24, 1965, at Stamford, Conn., Science and Technology Library, University of Akron, Akron, Ohio.

15. Morris, *American Synthetic Rubber Research Program*, 16.

16. Susan Calvo, "James D'Ianni: My Education and Career," unpublished compilation of papers, 2005, 139.

17. Morris, *American Synthetic Rubber Research Program*, 19.

18. Carl S. Marvel, tape-recorded interview by Herbert A. Endres, February 27, 1965, at the University of Arizona, Tucson, Science and Technology Library, University of Akron, Akron, Ohio.

19. Vernon Herbert and Attilio Bisio, *Synthetic Rubber: A Project that Had to Succeed* (Westport, Conn.: Greenwood Press, 1985), 152.

20. Ibid.

21. Ibid, 160.

22. In 1963, Ziegler and Natta shared the Nobel Prize in chemistry. The introduction of EPDM in the 1960s led to an entire new class of materials called thermoplastic elastomers (TPEs). TPEs combine the best of both worlds. TPEs process easily like plastics but retain their desirable rubber-like (elastomeric) properties. TPEs continue to be an important part of the rubber and polymer industries.

23. Ibid., 168.

24. Ibid. 160.

25. Morris, *American Synthetic Rubber Research Program*, 18.

26. Ibid., 22.

27. As division chairman in 1950, Stavely was instrumental in convincing the membership to raise money to pay the expenses of scientists from overseas

to attend division meetings. That year there were papers from England, France, Italy, and Germany.

28. The institute was later called the Institute of Polymer Science, and in 1993, the Maurice Morton Institute of Polymer Science.

29. Barbara Zimmerman, ed., *Vignettes from the International Rubber Science Hall of Fame (1958–1988): 36 Major Contributors to Rubber Science* (Akron, Ohio: Rubber Division of the American Chemical Society, 1989), 5.

30. David Giffels and Steve Love, "The Changing of the Guard," in *Wheels of Fortune: The Story of Rubber in Akron*, by David Giffels and Steve Love (Akron, Ohio: University of Akron Press, 1999), 321.

CHAPTER 12. CAPITAL INVESTMENT

1. Mister Wizard was the television name of Donald Herbert, a Wisconsin native who developed a fun way to introduce science to young people. A former World War II pilot, Herbert began his educational quest in 1949 at a Chicago NBC station. His last project was *Mr. Wizard's World*, airing three times a week on the Nickelodeon cable television network. He died of multiple myeloma on June 12, 2007. He was eighty-nine.

2. Junto Society, "About Junto Society," http://www.juntosociety.com/about.html/.

3. *Tel-Buch*, University of Akron Archives, 1925.

4. Josephine A. Cushman, *A Special Library for the Rubber Industry* (Akron, Ohio: Municipal University of Akron, 1920).

5. Gibbons interview by Endres.

6. Jack W. Neely, "The Development of the Library of the Division of Rubber Chemistry of the American Chemical Society" (master's thesis, Kent State University, 1955), 11.

7. Benjamin S. Garvey Jr., "The Rubber Library at the University of Akron Sponsored by the Division of Rubber Chemistry of the American Chemical Society: History," unpublished notes from Rubber Division meeting minutes, April 8, 1949, 1.

8. Benjamin S. Garvey Jr., "Report of the Committee on a Rubber Division Library," unpublished notes from Rubber Division meeting minutes, March 8, 1947, 1.

9. The *Bibliography of Rubber Literature* was first compiled and published by *Rubber Age* magazine in 1936. The division began compiling the fifth edition in 1940–41, but it was not published until after World War II.

10. Benjamin S. Garvey Jr. to the Committee on a Rubber Division Library, Rubber Division meeting minutes, November 10, 1947.

11. Benjamin S. Garvey Jr., "Report to the Directors of the Division of Rubber Chemistry of the American Chemical Society by the Committee on a

Library of the Division of Rubber Chemistry," unpublished notes from Rubber Division meeting minutes, October 27, 1948, 3.

12. Ibid., 4.

13. Ralph F. Wolf, "The Library of the Division of Rubber Chemistry, ACS," *Rubber World*, 1952, 125: 582–84.

14. Benjamin S. Garvey Jr., "Report to the Library Committee of the Division of Rubber Chemistry of the American Chemical Society," unpublished notes from Rubber Division meeting minutes, September 22, 1953, 1.

15. Dorothy Hamlen, Library Committee Meeting Minutes, Division of Rubber Chemistry, January 31, 1955, 1.

16. Ruth C. Murray, interview by author, August 2007, Cuyahoga Falls, Ohio.

17. Dorothy Hamlen et al., "Rubber Division Library," in *50th Anniversary of the Teaching of Rubber Chemistry* (Akron, Ohio: University of Akron Press, 1958), available at http://www2.uakron.edu/cpspe/polymer50/dvd/content/documents1958_20.html/.

18. Lois Brock, interview by author, August 2007, Akron, Ohio.

19. Panos Kokoropoulos, *Information Center for High Polymer Science and Technology* (Akron, Ohio: University of Akron Press, 1968), 43–46.

20. Panos Kokoropoulos and Sebastian V. Kanakkanatt, "Indexing System and Code for Polymers," *Journal of Chemical Documentation* (Akron, Ohio: University of Akron Press, 1968), 8(3): 179–87.

21. Joan Long, interview by author, July 2007, Bath, Ohio.

22. Robert Young, University of Akron and Bierce Library librarian, administered the Rubber Division library for a brief period of time in 1968.

23. Christopher J. Laursen, interview by author, January 2008, Akron, Ohio.

24. Jo Ann Calzonetti, interview by author, January 2008, Akron, Ohio.

25. Mark Bowles, interview by author, February 2008, Cuyahoga Falls, Ohio.

CHAPTER 13. THE COUNTDOWN

1. Mel Lerner, tape-recorded interview by Benjamin Kastein Jr., May 6, 1982, Science and Technology Library, University of Akron, Akron, Ohio.

2. The name of the committee was changed to Suppliers' Cooperative Reception in the mid 1960s.

3. Graff interview by author, August 2007.

4. D'Ianni interview by author.

5. John R. Deputy, telephone interview by author, March 2008.

6. Ed Noga, interview by author, March 2008, Akron, Ohio.

7. Scott et al., eds., "Division of Rubber Chemistry," October 1966, 69.

CHAPTER 14. METAMORPHOSIS

1. "Tire Industry FACTS, 2006," *Rubber Manufacturer's Association*, 2006.

2. Marge Bauer, tape-recorded interview by Benjamin Kastein Jr., August 26, 1986, Science and Technology Library, University of Akron, Akron, Ohio.

3. An internal promotion, titled "5,000 in '85," was conducted in 1985. Officially, records indicate the division had a total of 4,852 members after the completion of the promotion.

4. Rudy School, interview by author, November 2007, Cuyahoga Falls, Ohio.

5. Bauer interview by Kastein; Marge Bauer, interview by author, January 2008, Akron, Ohio.

6. Maurice Morton, *Introduction to Rubber Technology* (New York: Van Nostrand Reinhold, 1959).

7. Bauer interview by Kastein.

8. Steering committee meeting notes, November 9, 1961.

9. Steering committee meeting notes, April 24, 1962.

10. Steering committee meeting notes, October 16, 1962.

11. Today, there is no longer a Ph.D. in polymer chemistry per se, but two doctoral degrees are now available in polymer science and polymer engineering.

12. Robert Pett, telephone interview by author, January 30, 2008.

13. Howard L. Stephens, interview by author, January 2008, Akron, Ohio.

14. Frank N. Kelley, interview by author, June 2007, Hudson, Ohio.

15. "Maurice Morton Era," in *Pictorial History of Polymer Science at the University of Akron*, University of Akron, College of Polymer Science, http://www2.uakron.edu/cpspe/polymer50/dvd/content/images_69.html/.

16. Constance Morrison, interview by author, January 2008, Akron, Ohio.

17. Pett interview by author.

18. Robert E. Mercer, e-mail interview by author, August 2007.

19. Chris Probasco, telephone interview by author, December 2007.

20. Kurt Nygaard, telephone interview by author, December 2007.

21. Donald Mackey, telephone interview by author, January 2008.

22. Mark Petras, telephone interview by author, December 2007.

23. Toms Royal, e-mail interview by author, December 2007.

24. D'Ianni was chairman of the Rubber Division in 1964 and president of the American Chemical Society in 1980; W. J. Sparks was chairman of the Rubber Division in 1960 and president of the ACS in 1966; Harry L. Fisher was chairman of the Rubber Division in 1928 and president of the ACS in 1954.

25. Ralph Graff, telephone interview by author, February 2008.

26. Kent Marsden, interview by author, January 2008, Akron, Ohio.

27. Miller interview by author, September 2007.

CHAPTER 15. NURTURING HISTORY

1. David Morton, "Recording History," in *The History of Recording Technology*, 2008, http://www.recording-history.org/HTML/wire4.php/.

2. Gerald L. Baliles, Miller Center of Public Affairs, University of Virginia, Charlottesville, Virginia, http://millercenter.org/academic/presidentialrecordings/.

3. Benjamin Kastein, interview by author, February 2008, Silver Lake, Ohio.

4. Fran O'Connor, telephone interview by author, February 2008.

5. Historian Ben Kastein said the reality is that this objective was never a top priority, but the focus always remained on significant contributors to the rubber industry.

6. Leland Endres, telephone interview by author, February 2008.

7. Shelby Washko, interview by author, January 2008, Hinckley, Ohio.

CHAPTER 16. BORN OF TRAGEDY

1. Charles Rader, interview by author, February 2008, Akron, Ohio.

2. Keith Walker, telephone interview by author, February 2008.

3. Ed Miller, interview by author, February 2008, Akron, Ohio.

4. Ohio Rubber Group, "Scholarship," *Scholarship Update*, 2008, http://www.ohiorubbergroup.org/public/scholarship.cfm/.

CHAPTER 17. BACK INSIDE THE BOX

1. Miller interview by author, September 2007.

2. School interview by author.

3. Marsden interview by author.

4. Sue Barr, interview by author, January 2008, Akron, Ohio.

5. Scott et al., eds., "Division of Rubber Chemistry," October 1966, 74.

6. Ralph Graff, telephone interview by author, January 2008.

7. Kastein, "Historical Highlights of the Rubber Division ACS," 42.

8. D'Ianni interview by author.

9. This was the status of the ACS structure on June 11, 2007, the date of the interview.

10. Charles Rader, interview by author, September 2007, Akron, Ohio.

11. Barbara (Hodsdon) Ullyot, telephone interview by author, February 2008.

12. Calvo, "James D'Ianni."

13. Madeleine Jacobs, telephone and e-mail interview by author, February 2008.

INDEX

Vacca, George, 152, 216
van der Linde, Harold, 27
Vandenberg, Edwin J., 187, 220
Vargo, Richard, 195, 242
"The Variability of Crude Rubber"
  (Tuttle), 43
Vila, George M., 124
Virginia State University, 240
Vogt, Walter W., 87, 149, 215
vulcanization, 8, 22, 29, 49, 50, 69, 81, 83,
  165, 219; centennial celebration of, 81–
  82, 104
"Vulcanization of Rubber with Sulfur"
  (I. Williams), 87

Wai Sin Tiew, 48, 267n4
Walker, Andrew J., 142
Walker, Keith, 193, 194, 273n2
War Damage Corporation, 97
Warner, Walt, 155, 156, 226, 259
Warrick, Earl, 220
Washington Naval Treaty (1922), 77
Washko, Shelby, vii, P-12, P-19, P-21, 142,
  182, 186–187, 188–89, 190, 273n7
Waters, C. E., 27
Weber, Carl O., 130, 230
Weber, Lothar E., 86, 87
Weigand, William B., 78, 82, 106, 182, 221
Weigel, Kristine L., 142, 170
West Michigan Rubber Group, 54, 172,
  196
Whitby, G. Stafford, xvii, 46, 78, P-11,
  107–108, 111, 122–23, 130, 139, 164, 182,
  221, 224, 230, 257, 266n8, 268n22;
  contributions to the development of
  synthetic rubber, 107–9
White, Gaylon, xi, xiii,
White, James L., 219
Wickham, Henry, 9
Williams, Charles Hanson Greville, 69,
  130

Williams, Gay, 204
Williams, Ira, 61, 64, 78, 87, 215, 221
Williams, Robert R., 106, 107, 109
Willis, Graham P., xiii
Wilson, Woodrow, 41
Winkelman, Herbert A., 46, 53, 149, 150,
  214, 221
Wolf, Ralph S., 135, 163, 227
Wolff, Siegfried, 220
women, in the workforce, 119
Woodruff, William, 36, 262n1
Work, Bertram G., 38
World Intellectual Property Organization
  (WIPO), 76, 77
World War I, 36, 44–45, 74, 122; late entry
  of the United States into, 42–43;
  methyl rubber production during, 37
World War II, 77, 78, 79, 86, 113, 119, 133;
  attack on Pearl Harbor, 79; interest of
  the War Department in guayule, 84–
  85; military needs for rubber, 92;
  production figures for synthetic rubber
  during, 113; "Victory Program," 97;
  War Department budgets, 97. See also
  Synthetic Rubber Program
Wright, Bill, 176
Wright, Mary, 176
Wyrough and Loser, 175–76, 216, 217, 257

Yale University, 98, 121
Yokohama University, 94
Young, Robert, 271n22

Zelinski, Robert, 126
Zesiger, Denton, 162
Ziegler catalyst, 127
Ziegler, Karl, 127, 269n22
Zielasko, Ernie, xvi, P-16, 154–55, 156,
  187, 227
zinc, 25
zinc oxide, 29

The Rubber Division, ACS, wishes to thank the following sponsors for their generous contributions:

Centennial Elite Sponsors

BUCKINGHAM
DOOLITTLE & BURROUGHS, LLP
Attorneys & Counselors at Law
www.bdblaw.com

Polymer Valley
CHEMICALS, INC

.The.
University
of Akron

Centennial Book Supporter

R.D. Abbott Company, Inc.